Quality Handbook
for
Composite Materials

To my mother ...

... in memory.

Quality Handbook for Composite Materials

M. H. Geier

Lorient University
Lorient, France

and

Engineering School of the Ministry of Defence
France

CHAPMAN & HALL

London · Glasgow · Weinheim · New York · Tokyo · Melbourne · Madras

Published by Chapman & Hall, 2–6 Boundary Row, London SE1 8HN, UK

Chapman & Hall, 2–6 Boundary Row, London SE1 8HN, UK

Blackie Academic & Professional, Wester Cleddens Road, Bishopbriggs, Glasgow G64 2NZ, UK

Chapman & Hall GmbH, Pappelallee 3, 69469 Weinheim, Germany

Chapman & Hall Inc., One Penn Plaza, 41st Floor, New York NY 10119, USA

Chapman & Hall Japan, Thomson Publishing Japan, Hirakawacho Nemoto Building, 6F, 1-7-11 Hirakawa-cho, Chiyoda-ku, Tokyo 102, Japan

Chapman & Hall Australia, Thomas Nelson Australia, 102 Dodds Street, South Melbourne, Victoria 3205, Australia

Chapman & Hall India, R. Seshadri, 32 Second Main Road, CIT East, Madras 600 035, India

First English language edition 1994

© 1994 Technique et Documentation – Lavoisier

Original French language edition – *Manuel Qualité des Composites* – © 1989, Technique et Documentation – Lavoisier

Typeset in 10/12pt. Times by Thomson Press (India) Ltd, New Delhi

Printed in Great Britain at the University Press, Cambridge

ISBN 0 412 43120 3

A catalogue record for this book is available from the British Library

Library of Congress Catalog Card Number: 93–74883

♾ Printed on permanent acid-free text paper, manufactured in accordance with ANSI/NISO Z 39.48-1992 and ANSI/NISO Z 39.48-1984 (Permanence of paper).

Contents

Foreword

Professor Michel H. Geier's books are well-known in France where he is an authority on composites: raw materials (fibres or resins) as well as their classification, or testing and standards. Being thoroughly familiar with all these new materials the author knows perfectly well how to classify them and define the tests necessary to obtain high-quality products. At a time when standardization has become a science in itself, engineers, professors, students and manufacturers need a guide book which is both reliable and easy to consult.

Today several countries are trying to impose the standards developed by various organizations, both official and private, depending on the most developed segments of their main industry such as aerospace for the US; carbon fibres for Japan; and high technology for Europe.

The standard cannot be dissociated from a product and it constitutes the essential tool for marketing strategies. It plays a major economic role and is necessary to highlight the quality of a product.

Quality is Michel H. Geier's keyword. He shows that it can be defined vigorously so that all specialists can agree on the same definition, and he covers the whole field of composites with organic matrices; his book is indispensable for all those who work with composites.

Jean Marechal
President
Centre de Promotion des Composites (Paris)

Acknowledgements

The author extends his most grateful thanks to the companies, bodies and institutions that kindly contributed the brochures, literature and illustrations which were a great in the writing of this book. Among these are:

- l'Aérospatiale (Etablissement des Mureaux)
- l'AFNOR
- AIREX
- ARDROX SA
- l'ASTM
- Ciba-Geigy
- CDK-Composites
- Coremat
- Etablissements Cotton Frères
- Du Pont De Nemours France SA
- GIRA Ste
- INSTRON SA France
- Kleber-Colombes (Klegecell)
- Lemmens
- Electronika
- METRAVIB
- Mettler
- OSI
- Prodemat
- Rezolin
- Rhone-Poulenc
- STCAN (Departement MSN)
- Vetrotex

And special thanks to Madame Feuille, engineer in charge of composite standardization at AFNOR (France).

General points | 1

1.1 IDEAS OF QUALITY

1.1.1 Introductory notes

The blistering of the gel coat of 850 French-built yachts due to osmosis in sea water was widely debated in the French press in 1987. The gel coat had been catalysed with peroxide containing glycol. The shipyard sued the company that had supplied the product. On 3 February 1988, the Court of Appeal in Poitiers laid 25% of the blame on the shipyard for failing to analyse the catalyst before using it. In so doing, the courts emphasized once again the necessity for the manufacturer or the user to control the quality of all manufactured products.

The decision reached by the Poitiers court made this a test-case. The manufacturing of composite materials must, if done as seriously as that of other products, involve quality control of the following:

- supplies and raw materials;
- soundness of the laminate;
- mechanical, physical and chemical characteristics of the finished product;
- the skin-core adhesion of sandwich structures;
- dimensions of the finished product.

Quality control must be all the more systematic as:

- suppliers of the resin and reinforcing fibres are numerous;
- the quantities of products supplied are small;
- the making of composites is usually done by hand and by craftsmen. The handling of the materials is the source of random defects, the number of which depends on the experience and skill of the operator.

1.1.2 The aim of quality control

The main purpose of quality control is to satisfy the customer. Quality control will check that performance is in conformity with:

- that advertised by the manufacturer; or

- that specified in an agreement signed by both the manufacturer and the buyer, stating the price and delivery time.

The purpose is to meet the stated needs of the buyer; no more, no less. Exceeding these requirements would mean greater quality but at a higher cost either to the supplier or buyer. Failing to achieve it would mean insufficient quality.

A good product is therefore not synonymous with a high-performance one. In fact the following ratio must be specified:

$$\frac{\text{Technical performance}}{\text{Price} \times \text{Delivery time} \times \text{Lifetime}}$$

This manual will deal mainly with technical quality, for example the meeting of performance upon delivery and the maintenance of this performance during the 'profitable' life of the product.

1.1.3 Quality control: a definition

Quality control amounts to technical control of:

- all the stages of the manufacturing process;
- all raw materials (and possibly suppliers of these);
- all products whether they be subassemblies or full assemblies.

This control can take place:

- at the raw materials suppliers (upstream control);
- at the subcontractors (subassembly control);
- at the site in charge of the design, development and production of the finished product (assembly control).

Final assembly sometimes takes place at the client's premises. In this case, assembly control will be carried out there, against a reference schedule of conditions.

Quality is the concern of everyone who takes part in the making of the finished product, wherever they may be. One can thus say that quality is not something that is controlled but rather developed. Those responsible for quality control must be independent of manufacturers and must therefore have the power to reject products that do not comply with the requirements specified in the schedule of conditions. They must also have the capacity to impose modifications affecting manufacturing processes with a view to achieving the appropriate quality standards.

1.1.4 Quality assurance: a definition

Quality assurance is a wider concept than quality control. Quality assurance is the body of measures which guarantee in advance that the customer will be satisfied both from a technical and economic point of view. This concept, which is a relatively new one, emerged in the USA around 1960. In fact, the aim is to prove to the client beforehand that the product supplied to him will be satisfactory. This goal is achieved by carrying out quality control, but also by using all the methods which may lead to the manufacture of quality products, namely by:

- rules agreed with the customer and listed in a quality manual;
- ensuring that all raw materials are of a suitable standard;
- checking that such rules are heeded in the workshop under all circumstances;
- the filing of technical data, of data concerning the different stages of the manufacturing process, and of the results produced by quality controls at all levels.

When one is engaged in quality control, one is fighting against all the faulty and defective products which are put on the market. All the products that fail to comply with the schedule of conditions will therefore be eliminated when the quality controls have been completed.

The purpose of quality assurance is to eliminate the causes leading to the production of items that do not meet the buyer's requirements, that is, getting it right first time. A scheme of quality assurance (QA) is thus worked out:

- for a given product;
- on the shop floor;
- within the company.

The quality assurance system must be capable of:

- detecting the reasons why the product fails to comply with quality requirements after the completion of tests;
- suggesting remedies within the framework of a quality scheme with a view to turning out a satisfactory product in 100% of cases;
- verifying that the quality scheme agreed upon is being correctly implemented.

When quality assurance is well organized:

- the customer is satisfied; and
- manufacturing costs decrease because rejected products are virtually non-existent.

1.2 CLASSIFICATION OF QUALITY CONTROLS

These controls aim at:

- checking that supplies and raw materials comply with the schedule of conditions agreed upon in advance by both the suppliers and the users of the raw materials;
- periodically monitoring the quality of manufacturing.

Quality controls will tend to become less numerous and simpler as new standards are introduced and verified, as larger quantities of composites are produced, and as mastery of the different parameters involved in manufacturing processes improves.

However, in view of the current increase in the use of composites and the lack of comprehensive standards, controls must be systematically carried out at every stage of the manufacturing process.

To illustrate this, recall the scheme for the manufacturing of a 'contact' laminate through the so-called 'wet', process, as shown in Figure 1.1.

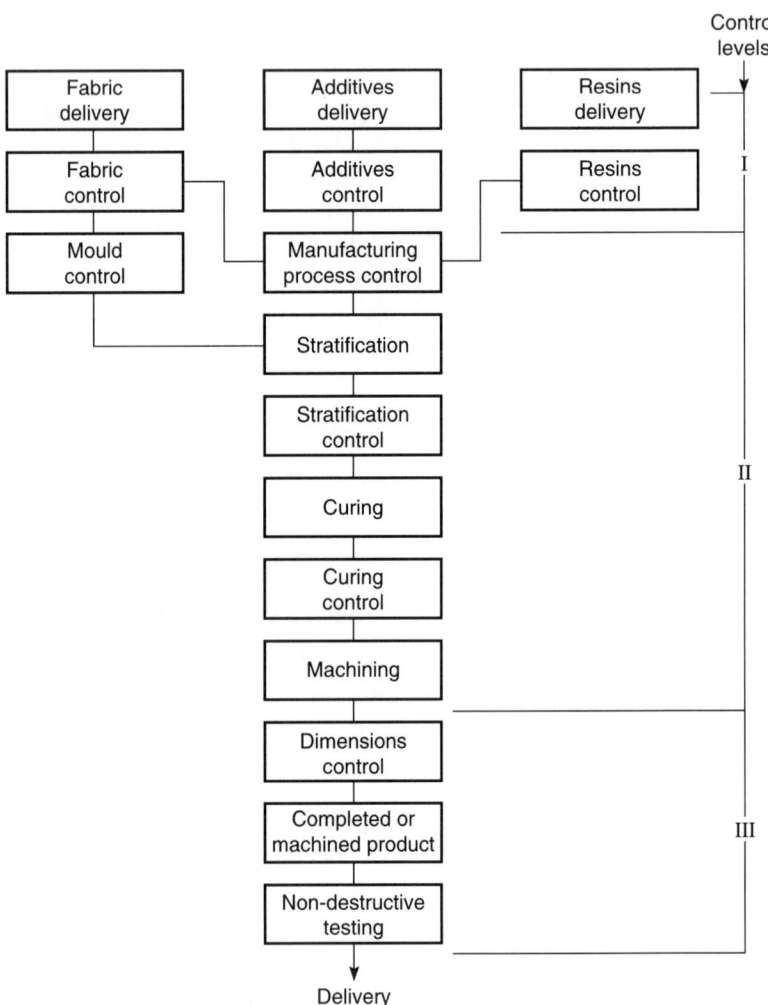

Figure 1.1 Scheme for the manufacturing of laminate through the 'wet' process.

Quality controls are carried out in the workshops at three separate levels:

Level 1: Quality control of raw materials as soon as they are delivered: e.g. fabrics, resins, gel coat hardeners, solvents and release agents, plus control of storage conditions.

Level 2: Quality control of the mould: dimensions, surface condition, controls of the different stages of manufacturing: laminating, curing, etc.

Level 3: Quality control of the manufactured composite product itself.

These controls, if they are to be systematic, must be cheap and easy to perform. Some controls are destructive, others are not (non-destructive testing: NDT). The latter tends to require sophisticated equipment and skilled manpower and consequently is more expensive. Only larger companies will use NDT and then only for products that require a high level of reliability: for example pressures vessels, and aeronautical and aerospace products.

1.3 STANDARDS

In France the Association Française de Normalisation (AFNOR) which has a very close working relationship with the International Standards Organization (ISO) and the Comité Européen de Normalisation (CEN), has published about 100 standards bearing on composite materials and their components. A collection of these standards has been published in volumes 1 and 2 of *Recueil des Normes Françaises*, 1991.

Volume 8 of *Recueil des Plastiques* (1992) deals with basic components for alveolar products and can be used for rigid foams which make up the core of sandwich composites.

Volume 1 of *Recueil Composites* (1991) deals essentially with glass fibre and carbon fibre reinforced composites.

Volume 2 of *Recueil Composites* (first edn) contains standards for resins used for matrices in composite manufacturing. Preimpregnated materials and finished products are also dealt with in this volume.

These volumes were published in 1986 and 1991. The officials now in charge of establishing standards are working on international standards (ISO) and European standards (CEN) to complete the collection concerning composite materials.

The person in charge of establishing standards for composite materials at AFNOR (Basic Industries department) is a specialist; she is a chemical engineer and she is currently developing a three-year standardization scheme for 1992–4.

AFNOR standards are technically in compliance with those of ISO. France is organizing work in the field at international level (ISO/TC61/SC13).

Nearly all NF (Norme Française) standards are based on ISO research. A table of relationships between French standards and ISO standards, as well as other main standards can be found at the end of volumes 1 and 2 of French standards (*Recueil des Normes Françaises*, 1991): see the list at the end of this chapter.

AFNOR and ISO standards for composites deal essentially with:

- specifications for fibres, reinforcement fabrics, preimpregnated fabrics and dry woven goods;
- test methods for characterization of glass, aramid and carbon fibres, unsaturated polyester and epoxy resins, preimpregnated fabrics and composite materials.

These test methods are of various kinds including chemical, physical and mechanical.

A European programme is now being prepared on standards for composite materials. As can be seen from the enclosed lists or from volumes 1 and 2 of *Recueil des Normes Françaises* (1991), published standards do not deal with quality control of laminates and composite sandwiches or non-destructive testing (NDT), but ISO is now turning its attention to these areas.

Composites are the main concern of a working group which is controlling a project called VAMAS (Versailles project on advanced materials and standards). This working group has the task of suggesting new standards for composites which will be agreed by ISO, AFNOR, ASTM, DIN, BS, CEN, etc., where:

- ISO: International Standards Organization
- CEN: Comité Européen de Normalisation
- AFNOR: Association Française de Normalisation
- ASTM: American Standards for Testing and Materials
- DIN: Deutch Industrial Normen
- BS: British Standards
- JIS: Japanese Standards.

The main addresses are:

- AFNOR: Association Française de Normalisation, Tour Europe, 92080 Paris – la Défense – Cedex 7, France.
- ASTM: American Standards for Testing and Materials, 1916 Race Street, Philadelphia, PA 19103-1187, USA.
- DIN: Deutch Industrial Normen, Burggrafenstrasse 6, Postfach 1107, Deutch 1000, Berlin 30, Germany.
- BSI: British Standards Institution, 2 Park Street, London, W1A 2BS, United Kingdom.
- JISC: Japanese Industrial Standards Committee, CO Standards Department, Agency of Industrial Science and Technology, Ministry of International Trade and Industry, 1-3-1 Kasumigaseki, Chiyoda-ku, Tokyo 100, Japan.

As far as CEN is concerned, technical committee 249 is in charge of defining standards. The CEN/TC249/SC1 subcommittee, which deals with resins, is presided over by the United Kingdom and has its secretariat at the BSI in London. CEN/TC249/SC2 subcommittee 2 deals with fibres, rovings, prepregs and composites. It is presided over by France and has its secretariat at AFNOR in Paris.

Since 1990, these committees have been busy updating the standards for composite materials. CEN's Technical Bureau, which is the executive body, will approve the programmes of both TC249 SC1 and SC2 in 1992.

As far as ISO is concerned, technical committee 61 is in charge of defining standards. ISO/TC61/SC12 subcommittee 12 deals with resins. ISO/TC61/SC13 subcommittee 13 deals with fibres, rovings and prepregs. It is presided over by France and has its secretariat at AFNOR in Paris. The updating of standards is also in progress there. The purpose is to develop test standards at ISO and to use these methods for CEN. CEN standards are used by ISO for establishing product specifications. Generally new standards produced by AFNOR, ISO and CEN are now harmonized.

The European producers and suppliers of aerospace materials have united within a federation called AECMA which works on standards for composite materials within the Comité Européen de Normalisation (CEN). A similar body known as SACMA is also active in the USA. To date AECMA has produced the following standards: EN 2561, EN 2562, EN 2563 and EN 2377 for CEN. The following appendices give the scheduled activities of AECMA under the eurostandard project (prEN).

APPENDIX A   French standards (1992)

A.1 Glossary

Textile glass	NF ISO 6355 – NF ISO 2078
Prepregs – symbols – definitions	NF ISO 8604

A.2 Sampling results

Glass fibre reinforced plastics (GRP): preparation of test specimens	NFT 57-151, NFT 57-152, NFT 57-153
Carbon fibre reinforced composites (prepregs): manufacture of test pieces	NFT 57-300
Statistical precision of test methods	NF ISO 5725
Statistical interpretation of test results	NF X 06-042
Miscellaneous – machining of test specimens	NFT 58001

Miscellaneous – tolerances of moulded parts	NFT 58000
Miscellaneous – preparation of test specimens	NFT 58004
Sampling methods for checking lots	NFT 25001
Compression moulding test specimens from powder granules or paste	NFT 58005

A.3 Tests

(a) Reinforcing fibre

Glass fibre

Combustible matter content	NFB 38-101
Average diameter of fibre	NFB 38-102
Linear density of yarns	NF ISO 1889, NFB 38-105
Twisting index of yarns	NF ISO 1890, NFB 38-109
Water content or moisture ratio	NFB 38-108
Width and length of woven fabrics	NFB 38-204
Tensile test on yarns	NF ISO 3341
Roving stiffness	NFB 38-152
Tensile tests on impregnated rovings	NFB 38-153
Slip resistance of threads in woven fabrics	NFB 38-202
Tensile test on woven fabrics	NFB 38-203
Flexural stiffness of woven fabrics	NFB 38-206
Dissolution time of binder in styrene	NFB 38-302
Mats, average thickness	NFB 38-303
Mass per surface area of strand mat	NFB 38-304
Tensile test on strand mat	NF ISO 3342

Carbon and graphite fibres

Fibres, woven fabrics: visual defects	NFT 25-002
Fibres (yarns, rovings): specific gravity	NFT 25-100
Impregnated fibres: tensile test	NFT 25-101
Linear weight of fibres	NFT 25-103
Size ratio of fibres	NFT 25-104
Diameter of fibres	NFT 25-106
Chemical analysis	NFT 25-107, NFT 25-112

Aramid fibres

Fibres, woven fabrics: visual defects	NFT 25-002
Size ratio of fibres	NFT 25-401
Linear weight of fibres	NFT 25-402
Water content of fibres	NFT 25-403

Ceramic fibres
Linear weight of fibres   NFT 25-701
Size ratio of fibres   NFT 25-702

(b) Preimpregnated fibres

Prepregs, glass mats, yarns, rovings, SMC, BMC

Plasticity	NFT 57-513
Mouldability	NFT 57-514
Measurement of expansion	NFT 57-515
Conventional reactivity	NFT 57-516
Glass and filler content: burn-off method	NFT 57-518
Shrinkage during compression casting	NFT 57-519
Conventional volatile matter content	NFT 57-570
Determination of ignition loss	NFT 57-571
Conventional resin flow	NFT 57-603
Fibre and resin content: dissolution method	NFT 57-608

Preimpregnated fabrics (prepregs)

Thickness of layer after pressing and curing	NFT 57-556
Binding time	NFT 57-559
Mass per unit area	NFT 57-601
Conventional volatile matter	NFT 57-602
Conventional resin flow	NFT 57-603
Fibre content: dissolution method	NFT 57-608
Glass and filler content: burn-off method	NFT 57-518
Determination of loss through calcination	NFT 57-557
Differential scanning calorimetry	NFL 17-451

Miscellaneous (BMC)

Compression moulding test: powder, granule, bulk compound	NFT 58-003
Thermoplastics injection moulding: determination of characteristics	NFT 58-007
Thermosetting plastics – injection moulding: determination of characteristics	NFT 58-008

(c) Resins

Tests for thermosetting resins

Phenolic resins

Phenolic powder – flow rate in melted condition	NFT 51-421
Phenolic powder – sieve analysis	NF ISO 8620

Liquid phenolic resin – water tolerance	NFT 51-423
Liquid phenolic resin – dry extract	NFT 51-425
Phenolic resins – density	NFT 51-426
Liquid phenolic resin – kinematic viscosity	NFT 51-427
Measurement of reactivity	NFT 51-428
Determination of gel time	NFT 51-429
Water content (Karl Fischer method)	NFT 51-430
HMTA content	NFT 51-431, NFT 51-432
Free formaldehyde content	NFT 51-434
Residual phenol: gas chromatography	NFT 51-436
Solid resin – softening point	NFT 51-437
Solid resin – non-volatile matter	NF ISO 8618
Summary of test methods	NFT 51-440
IR spectral analysis	NFT 51-500
Chromatography, GPC analysis	NFT 51-505
Chromatography, HPLC analysis	NFT 51-506

Unsaturated polyester resins

IR spectral analysis	NFT 51-500
Chromatography, GCP analysis	NFT 51-505
Chromatography, HPLC analysis	NFT 51-506
Volume shrinkage on curing	NFT 51-501
Determination of acid value	NFT 51-511
Gel time at 25 °C	NFT 51-512
Hydroxyl value	NFT 51-513
Conventional reactivity at 80 °C	NFT 51-514
Conventional content of volatile matter	NFT 51-515
Determination of properties	NFT 51-516
Viscosity under action of MgO	NFT 51-517
Conventional reactivity at 130 °C	NFT 51-518

Epoxies

Determination of epoxy equivalent	NFT 51-522
Inorganic chlorine by direct titration	NFT 51-523
Saponifiable chlorine	NFT 51-524
Conventional content of volatile matter	NFT 51-525
Tendency to crystallize	NFT 51-526
Determination of properties	NFT 51-527
DDA hardening agent content (IR spectrum)	NFT 51-528
Epoxide index (IR spectrophotometry)	NFT 51-529
Volume shrinkage on curing	NFT 51-501
IR spectroanalysis of resin	NFT 51-500
Chromatography, GPC analysis	NFT 51-505
Chromatography, HPLC analysis	NFT 51-506

Tests for thermoplastics and thermosetting resins

Deflection under a load	NFT 51-222
Temperature of deflection under load	NF ISO 75
Standard atmospheres for testing	NFT 51-014
VICAT softening temperature of thermoplastics	NFT 51-021
Resistance to water and chemical liquids	NFT 51-029
Heat deformation (Martens)	NFT 51-070
Bending fatigue test	NFT 51-119, NFT 51-120
Determination of filler content (fibre glass, MoS_2, carbon black)	NFT 51-143, NFT 51-144, NFT 51-146
Water absorption ratio	NFT 51-166
Liquid resin, density (pycnometer)	NF ISO 1676
Liquid resin, Brookfield viscosity	NFT 51-210
Liquid resin, kinematic viscosity	NFT 51-211
Shrinkage of thermosetting moulding materials	NFT 51-40
Transfer flow of thermosetting moulding resins	NFT 51-402

(d) Composites

Glass reinforced plastics (GRP)

Standard atmospheres for conditioning and testing	NFT 57-050
Visual defects	NFT 57-100
Tensile properties	NFT 57-101
Determination of loss on ignition	NFT 57-102
Compression test	NF ISO 8515
Delamination under bending (short beam)	NF ISO 4585
Flexural test	NFT 57-105
Hardness: Barcol test	NFT 57-106
Change of characteristics during hot water treatment	NFT 57-107
Impact strength (Charpy) test	NFT 57-108
Statistical cavities ratio	NFT 57-109
Apparent interlaminar shear strength	NF EN 2377
Perpendicular tensile test	NFL 17-452
Peel strength	NFL 17-455

Carbon fibre reinforced composites (CFRP)

Tensile test	NFT 57-301
Flexural test (three-point loading)	NFT 57-302
Delamination test under bending	NFT 57-303
Carbon fibre resin UD laminates – tensile test	NFL 17-410
Carbon fibre resin UD laminate – flexural test	NFL 17-411
Carbon fibre resin UD laminate – interlaminar shear test	NFL 17-412
Transverse tensile test	NFL 17-452
Peel strength	NFL 17-455

A.4 Designation

(a) Fibres

Textile glass fibre yarns	NF ISO 2078
Glass woven fabrics	NFB 38-205
Carbon fibre yarns	NFT 25-110
Carbon woven fabrics and webs	NFT 25-200

(b) Resins – additional components

Unsaturated polyester resins	NFT 51-510
Epoxy resins	NFT 51-520
Hardeners and accelerators for epoxy resins	NFT 51-521
Phenolic resins	NFT 51-420
Base polymer symbols	NF ISO 1043.1
Fillers – reinforcing material – symbols	NF ISO 1043.2
Plasticizer symbols	NF ISO 1043.3
Reinforced thermosetting plastics	NFT 57-000

A.5 Basis for specification

(a) Glass fibre

Textile glass: yarns	NF ISO 3598
Textile glass: rovings	NF ISO 2797
Textile glass: mats	NFB 38-301
Textile glass: special yarns	NF ISO 8516
Sheet moulding compound, SMC	NF ISO 8605

(b) Carbon fibre

Prepregs of carbon fibre woven fabrics	NFT 57-700
Carbon fibre woven fabrics and webs	NFT 25-200
Carbon fibre yarns	NFT 25-108, NFT 25-109

APPENDIX B   ISO standards – ISO TC61 position 1991; studies, projects, reviews

B.1 Work of ISO TC61/SC12: Resins

Compression moulding of thermosetting specimens: in review by ISO 295.
Resin ratio of phenoplast moulding material: in review by ISO 308.

Phenolic moulding: specifications for moulding material: in review by ISO
  800.
Phenolic resins – glossary.
Phenolic resins: viscosity, UBBELOHDE.
Phenolic resins: free formaldehyde content.
Phenolic resins: reactivity and melting point by DSC.
Phenolic resins: liquid chromatography analysis.
Unsaturated polyester: volatile matter content.
Unsaturated polyester: gel time at 25 °C.
Unsaturated polyester: acid value.
Unsaturated polyester: hydroxyl value.
Unsaturated polyester: designation.
Unsaturated polyester: characterization.
Unsaturated polyester: volume shrinkage.
Epoxies: volatile matter content.
Epoxies: epoxy equivalent.
Epoxies: inorganic chlorine.
Epoxies: volume shrinkage.
Epoxies: saponifiable chlorine.
Epoxies: designation.
Epoxies: characterization.
Epoxies: hardeners and accelerators.
Hardeners for epoxies: acid value.
Amine hardeners for epoxies: classification.
Thermosetting resins: miscellaneous, testing methods.
Glycidic ester resins: inorganic chlorine.
Thermosetting resins: IR spectroanalysis.

B.2 Work of ISO TC61/SC13: Reinforcements: fibres, woven fabrics, prepregs, SMC, BMC, DMC

GRP glass reinforced plastics, prepregs: fibre content and mineral fillers.
Textile glass: volatile matter content.
PRVT: methods for characterization.
GRP: deflection temperature under load.
PRVT: reinforced plastic textile glass: tensile test.
PRVT: impact strength (Charpy) test.
GRP: bending fatigue test.
Glass fibres: average diameter.
Glass fibre yarns: designation.
Glass fibre yarns: tensile test.
Glass fibre woven fabrics: thickness.
Glass fibre woven fabrics: tensile test.
Glass fibre woven fabrics: flexural stiffness.

Glass fibre woven fabrics: width and length.
Glass fibre woven fabrics: mass per unit surface area.
Mats: thickness under load.
Preimpregnated roving: tensile test.
Manufacture of test sheets by low pressure technique.
Manufacture of UD test pieces by filament winding.
Prepregs: glass content (dissolution method).
Moulding test for prepregs.
SMC, DMC, BMC: conventional reactivity.
Glass prepregs: glass content (burn-off method).
Storage life for prepregs.
Prepregs: volatile matter content.
Prepregs: mass per unit surface area.
Cross-linking of SMC and BMC.
Graphite fibres: designation.
Graphite and carbon fibres: vocabulary.
Carbon fibres: linear weight, density.
Graphite fibres: size ratio.
Preimpregnated carbon yarns: tensile test.
Graphite fibres: diameter of fibres.
Graphite rovings: tensile test.

APPENDIX C   AECMA schedule (CEN) 1991

C.1 Published standards (CEN)

UD laminate carbon fibre resin: tensile test	EN 2561
UD laminate carbon fibre resin: flexural test	EN 2562
Laminate carbon fibre resin: interlaminar shear test	EN 2563
Glass fibre reinforced plastics (GRP): interlaminar shear test	EN 2377

C.2 Standards in preparation (pr EN)

Glass fibre prepregs: mass per unit area	pr EN 2329
Glass fibre woven fabrics: content of volatile matter	pr EN 2330
Glass fibre prepregs: resin content	pr EN 2331
Glass fibre prepregs: resin flow	pr EN 2332
Laminate glass fibre resin, sandwich structures: manufacture of test sheets	pr EN 2374
Prepregs: range of lots	pr EN 2375
GRP, glass fibre reinforced plastics: water content	pr EN 2378
GRP, glass fibre reinforced plastics: chemical resistance	pr EN 2489
Carbon fibre prepregs: mass per surface area	pr EN 2557
Carbon fibre prepregs: conventional volatile matter	pr EN 2558

Carbon fibre prepregs: fibre and resin content	pr EN 2559
Carbon fibre prepregs: resin flow	pr EN 2560
Carbon fibre resin laminate: fibre and resin content, porosity	pr EN 2564
Carbon fibre resin laminate: manufacture of test sheets	pr EN 2565
UD laminate carbon fibre resin transverse tensile test	pr EN 2597
GRP: standard atmosphere for conditioning and testing	pr EN 2743
GRP: density	pr EN 2745
GRP: flexural test (three-point loading)	pr EN 2746
GRP: tensile test	pr EN 2747
GRP: resistance to hot moist conditions	pr EN 2823
Glass fibre prepregs: supply specifications	pr EN 2833
GRP: compression test	pr EN 2850
Carbon fibre yarns (polyacrylonitrile): specifications	pr EN 2962
Graphite yarns ($\sigma > 3000$ MPa, $E > 220$ GPa) specifications	pr EN 2963
Carbon fibre yarns: linear density	pr EN 2964
Carbon fibre yarns: average diameter of fibres	pr EN 2965
Carbon fibre yarns: size ratio	pr EN 2967
Carbon fibre yarns: tensile test (σ, E)	pr EN 3075
Carbon fibre yarns: density	pr EN 3076
GRP: moisture content	pr EN 3615, pr EN 3616

APPENDIX D   Equivalence between standards

Table 1.1   Control of fibres and dry woven fabrics

Title	NF	ISO	BS	DIN	ASTM	CEN
Glass fibre – combustible matter content, size content	B38-101	1887		52330	D 578	
Glass fibre – average diameter of filaments	B38-102	1888				
Glass fibre – yarns and rovings linear density	NF ISO 1889	1889		53830		
Glass fibre yarns – determination of twist	NF ISO 1890	1890				
Textile glass fibre – glass fabrics linear density of yarns	B38-105					
Roving glass fibre – yarn glass fibre, basis for a specification	NF ISO 3598	3598		60850	D 578	
Glass fibre: yarns, rovings, mats, fabrics: water content	B38-108	3344				

Table 1.1 (*continued*)

Title	NF	ISO	BS	DIN	ASTM	CEN
Glass fibre yarns: twisting index	B38-109	3343				
Glass fibre rovings and yarns: tensile test	NF ISO 3341	3341		53834T$_1$ part 1		
Glass fibre roving – basis for a specification	NF ISO 2797	2797				
Glass fibre roving – stiffness	B38-152	3375				
Glass fibre woven fabrics – tensile test (strip method)	B38-203	4606		53857		
Glass fibre – woven fabric width and length	B38-204	5025				
Glass fibre – woven fabric – basis for specification	B38-205	2113				
Glass fibre – woven fabric, conventional flexural stiffness	B38-206	4604				
Glass fibre mats – plan for a specification	B38-301	2559				
Chopped strand mat – dissolution of binder in styrene	B38-302	2559				
Glass fibre mat, average thickness	B38-303	3616		53855T$_1$ and T$_2$		
Glass fibre mats, mass per unit surface area	B38-304	3374		53854		
Glass fibre mats, tensile test	NF ISO 3342	3342		53857		
Glass fibre textured yarns, basis for a specification	NF ISO 8516	8516				
Conventional atmosphere for conditioning and testing of textiles	NFG 00-003	139	1051	50014 53820		
Textiles, linear density [tex] fabrics	NFG 01-001	1144				
Yarns, tensile test	NFG 07-003	2062				
Fabrics, surface density	NFG 07-150	3801		53854	3786	
Woven fabrics, average thickness	NFG 07-153	5084		53855	D 578	
Woven fabrics, mass of warp and weft per unit area	NFG 07-157	7211/6				
To count yarns or rovings in warp and weft	NFG 07-155	7211/2		53853	D 3775	
Carbon fibres – density	T25-100	DIN 10119				
Carbon fibre yarns, rovings – tensile test	T25-101	CD 10618				
Carbon fibres – size content	T25-104	CD 10548				

Table 1.1 *(continued)*

Title	NF	ISO	BS	DIN	ASTM	CEN
Carbon fibres – diameter of filament yarn	T25-106	CD 11567				
Carbon fibres – metal impurities	T25-107					
Carbon fibres – chemical analysis – metalloid elements	T25-112			65579		
Aramid fibres – size content	T25-401					
Aramid fibres – linear mass of yarns	T25-402					
Aramid fibres – water content	T25-403					
Aramid fibres – filament diameter	T25-405					
Ceramic fibres: linear mass	T25-701					
Ceramic fibres: size content	T25-702					
Ceramic fibres: tensile test on monofilament	T25-704					

Table 1.2 Control of resins

Title	NF	ISO	BS	DIN	ASTM	CEN
Liquid phenolic resin: water tolerance	T51-423	8989				
Phenolic resin: dry extract	T51-425					
Phenolic resin: density	T51-426	9371				
Phenolic resin: viscosity	T51-427					
Phenolic resin: reactivity	T51-428	9887				
Phenolic resin: gel time	T51-429	9396				
Phenolic resin: water content	T51-430					
Phenolic resin: HTMA content	T51-431 T51-432					
Phenolic resin: free formaldehyde content	T51-434	9397				
Thermosetting resins: IR spectral analysis	T51-500					
Thermosetting resins: CGP analysis	T51-505					
Thermosetting resins: HPLC analysis	T51-506					
Liquid resins: density pycnometer method	NF ISO 1676	1676				
Liquid resins: Brookfield viscosity	T51-210	2555				
Liquid resins: kinematic viscosity	T51-211	3219				

Table 1.2 (*continued*)

Title	NF	ISO	BS	DIN	ASTM	CEN
Unsaturated polyesters, epoxy resins: volume shrinkage	T51-501	3251	2782/6 644A	16945		
Unsaturated polyesters: acid value	T51-511	2114	2782/4 432B	53402		
Unsaturated polyesters: gel time at 25 °C	T51-512	2535	2782	16945		
Unsaturated polyesters: volatile matter content (styrene ratio)	T51-515	7028		16945		
Unsaturated polyesters: reactivity at 80 °C	T51-514	584		16945		
UP resin: viscosity, influence of MgO	T51-517					
UP resin: reactivity at 130 °C	T51-518					
UP resin: filler content by loss through calcination	T57-557					
UP resin: total chloride ratio		4615		53474		
UP resin: hydroxyl value	T51-513	2554	2782/4 432C	53240		
Resins: temperature of deflection under a load	NF ISO 75	75				
Resins: standard atmospheres for testing	T51-014					
Resins: deflection under load	T51-222					
VICAT softening temperature for thermoplastics	T51-021	291				
Thermosetting resins: shrinkage and post-shrinkage	T51-401	2577				
Thermosetting resins: transfer flow	T51-402					
Resins: impact test on cured resins	T51-035					
Resins: bending test on cured resins	T51-001				D 790	63
Resins: water absorption ratio	T51-166	62				
Resins: tensile test on cured resins	T51-034 T51-516	R527				
Epoxies: epoxy equivalent	T51-522	3001	2782/4 432C 433D	16945		
Epoxies: inorganic chlorine	T51-523	4573	2782/4 433A			
Epoxies: saponifiable chlorine	T51-524	4583	2782/4 433B			
Epoxies: content of volatile matter	T51-525	in prep.				

Table 1.2 (*continued*)

Title	NF	ISO	BS	DIN	ASTM	CEN
Epoxies: determination of properties	T51-527	in prep.			D 638	
Epoxies: tendency to crystallize	T51-526					
Epoxies: IR spectro-photometry epoxy index	T51-529					
Epoxies: IR spectro-photometry, DDA hardening agent content	T51-528					
Epoxies: spectroanalysis GPC	T51-505					
spectroanalysis HPLC	T51-506					

Table 1.3 Control of preimpregnated mats, SMC, BMC, DMC, and preimpregnated yarns, rovings

Title	NF	ISO	BS	DIN	ASTM	CEN
Glass mats, SMC, DMC weight per unit surface area	T57-511					
Glass fibre mats, SMC, DMC plasticity	T57-513					
Glass fibre mats, SMC, DMC: mouldability	T57-514	4900				
Glass fibre mats, SMC, DMC expansion test	T57-515					
Glass fibre mat, SMC, BMC, DMC: volatile substances content	T57-512					
Glass fibre mat, SMC, BMC, DMC: reactivity	T57-516	9780				
Glass fibre mat, SMC, BMC, DMC: glass and fillers content, burn-off method	T57-518	1172				
Mats, SMC: shrinkage during compression casting	T57-519					
Mats, SMC, DMC: conventional volatile matter content	T57-570					
Mats, SMC, DMC, yarns: glass content, ignition loss	T57-571					
Mats, SMC, DMC: fibre and resin content: dissolution method	T57-608					
BMC, DMC: compression moulding test	T58-003	293				

Table 1.4 Control of preimpregnated fabrics

Title	NF	ISO	BS	DIN	ASTM	CEN
Prepregs: fibre and resin content, dissolution method	T57-608					
Prepregs: glass-resin: glass fibre and filler content: burn-off method	T57-518					
Prepregs: differential scanning calorimetry	L17-451					
Expoxy prepreg: shear adhesion to copper	T57-555					
Thickness of prepreg layer after pressing and curing	T57-556					
Prepreg glass fibre fabric: loss through calcination	T57-557					
Glass epoxy prepreg: binding time	T57-559					
Prepreg: mass per unit area	T57-601	10352				
Prepreg: conventional volatile matter	T57-602	9782				
Prepreg: conventional resin flow	T57-603					

Table 1.5 Control of sandwich core

Title	NF	ISO	BS	DIN	ASTM	CEN
Rigid foam, shear test	T56-118	1922			D 4255	
Apparent density of foam	T56-107	845				
Moisture ratio and volatile matter in foam	T56-106	no standard				
Tensile test on rigid foam	T56-103	1926				
Bending test on rigid foam	T56-102	1209				
Punching test on rigid foam	T56-104	no standard				
Compression test on rigid foam	T56-101	844				
Foams: accelerated ageing test	T56-117	2440				
Conventional atmosphere (humidity, temperature) for testing plastics	T51-014 / T57-050	291	2782/10 1004	50014		EN62
Glass microballs: crushing or compression test					D 3101-72	

Table 1.6 Control of laminate: glass reinforced plastic (GRP)

Title	NF	ISO	BS	DIN	ASTM	CEN
Standard atmospheres for conditioning and testing	T57-050	291	2782/10 1004	50014		EN62
Visible defects	T57-100					
Tensile test	T57-101	3268	2782/10 1003	53455 392-457	D 3039 D 638	EN61
Loss on ignition: glass and resin values	T57-102	1172	2782/10 1002	53395		EN60
Laminated glass – UP resin ratio of residual styrene		4901 1985				
Flexural test (three-point method)	T57-105	178	2782/10 335A	53452	D 790 part 10	EN63
Compression test parallel to the plane of lay-up	NF ISO 8515	8515		53454	D 695	
Delamination under bending (short beam)	NF ISO 4585	4585			D 2344 D 2345	
Impact strength (Charpy) test	T57-108			53453	D 256	
Hardness (Barcol) test	T57-106	868				EN59
Pore ratio by statistical method	T57-109	7822				
Preparation of UD sheets for other tensile tests	T57-151 T57-152					
Preparation of low-pressure laminates for test purposes	T57-153	1268				
Apparent interlaminar shear strength	NF EN2377					EN 2377
Rail shear test					D 4255	
Shear using a notched specimen					E 399	
Coulomb's modulus				53445		

Table 1.7 Control of laminate: carbon fibre reinforced plastic (CFRP)

Title	NF	ISO	BS	DIN	ASTM	CEN
Prepregs, carbon resin: manufacture of test plates	T57-300					
Tensile test	T57-301	3268	2782/10 1003	53455 392-457	D 3039 D 638	EN61
Flexural test (three-point test)	T57-302	178	2782/10 335A	53452	D 790 part 10	EN63
Delamination under bending (short beam)	T57-303	4585			D 2344 D 2345	
Carbon fibre resin UD laminates tensile test	L17-410					Pr EN2561
Carbon fibre resin UD laminates interlaminar shear strength	L17-412				D 2344 D 2345	Pr EN2563
Carbon fibre resin UD laminates flexural test	L17-411	178	2782/10 335A	53452	D 790 part 10	Pr EN2562
Differential scanning calorimetry	L17-451					

Table 1.8 Bonding tests: control of sandwich structures

Title	NF	ISO	BS	DIN	ASTM	CEN
Perpendicular tension on a glued joint	L17-452	no standard				
Peeling test on the skin of a sandwich	L17-455 T76-112	no standard		53289	3167-76	

Table 1.9 Bonding test: control of adhesives

Title	NF	ISO	BS	DIN	ASTM	CEN
Determination of dry extract	T76-101				D 898-69	
Brookfield viscosity of liquid adhesives	T76-102				D 899-51	
Ignition loss (burn-off method)	T76-110					
Conventional reactivity	T76-116					
Listing of testing methods	T76-300 T76-303					
Impact strength, IZOD, test	T76-115				D 950-82 D 950-78 D 950-72	
Creep tests (a) long-term shear test	T76-119 T76-122			53284 55284	D 2294-69 D 1780-72	
(b) peel test					D 2918-71 D 2919-71	
Tensile test on polymerized adhesive	T51-034			53455		
Preparation of sheets of thermoset adhesive for test purposes	T76-142					

Table 1.10 Control of bonded joins

Title	NF	ISO	BS	DIN	ASTM	CEN
Single shear test	T76-107	4587		53283	D 1002-72	
Double shear test					D 3258-76	
Shear test curve (σ, ε)	T76-141			54451	D 3165-73	
Bending test					D 1184-69	
Test under shear	T76-111	9664			D 3166-76	
Fracture (Boeing) test	T76-114				D 3762-79	
Fracture strength DCB test					D 897 -72 D 1062-78 D 3807-79	

Control of materials | 2

2.1 CONTROL OF FIBRES AND DRY WOVEN FABRICS

Fibres and reinforcements come in the form of tows, rovings, dry woven fabrics or non-woven fabrics or prepregs. Fabrics and dry strands or tows currently account for most of reinforcement consumption although the use of preimpregnated material is growing. The objective is to control the following parameters:

- the mechanical properties of the reinforcement;
- the ability of the fibre to bond to the resin matrix.

Verification of these properties involves the use of standard procedures, which have so far been employed for glass fibre reinforcement (1990).

2.1.1 Main standards

NFB 38-110 or ISO 3341. Determining the rupture strength and rupture elongation during strand tensile tests.

NFB 38-203, ISO 4606 or DIN 58857. Determining the rupture strength and rupture elongation during fabric strip tensile tests.

NFB 38-304, ISO 3374 or DIN 53854. Determining the weight per unit area of glass mats.

NFG 07-104 or ISO 4605. Determining the weight per unit area of glass fabrics.

NFG 07-150, ISO 3801, ASTM 5786 or DIN 53854. Determining the weight per unit area of knitted and woven fabrics.

NFB 38-103, ISO 1889 or DIN 53830. Determining the linear weight of glass strands.

NFT 25-107. Chemical analysis of fibre carbon.

NFT 25-100 or ISO 10119. Determining the volumetric weight of carbon fibre.

NFT 25-104 or ISO CD 10548. Determining the amount of size on carbon fibres.

Specification of glass fibre yarns and rovings: ISO 3598, NFB 38-107, DIN 60850 or ASTM D 578.

Specification of carbon fibre yarns and rovings: NFT 25-108.

Specification of glass fibre fabrics: ISO 2113, NFB 38-205 or DIN 61854 and DIN 16740, ASTM D 579 or BS 3396 and 3747.

NFB 38-101, ISO 1887, DIN 52330 or ASTM D 578. Determination of combustible matter in glass fibre yarns and rovings (this is used to assess the organic size content of glass fibre).

NFG 00-003, ISO 139, BS 1051 or DIN 50014 and DIN 53820: normal humidity and temperature for conditioning fabrics, rovings or yarns before testing.

2.1.2 Visual control of woven fabric

This is the first control to be carried out upon taking delivery of material. The fabric must be:

- of the type ordered (web, serge, satin, etc.) and, in the case of satin (4, 5 or 8 harness, etc.);
- even in structure; weaving defects can generate flaws in the composite due to entrapment of surplus resin delamination, etc;
- of constant width; conformity with order (e.g. ± 3%).

2.1.3 Control of weight per unit surface area

The specimens (yarns, rovings, fabrics) are conditioned in advance in an oven or enclosure for at least six hours (temperature 23 °C, 50% relative humidity) to obtain a constant weight. Standards: ISO 139, NFG 00-003, BS 1051 or DIN 50014 and 53820.

(a) Fabrics and mats

Samples measuring $S(m^2)$ are cut off and weighed accurately to calculate the mass M (g), from which the mass per unit area can be found. The average mass per unit area, m, is based on five specimens with a tolerance of ± 10%

$$m = \frac{M}{S} \quad (g/m^2).$$

Standards: ISO 3801, ISO 4605 or ISO 3374; NFG 07-150 or NFB 38-304; DIN 53854; ASTM 3786.

(b) Rovings or multiple wound yarns

In the case of yarns and rovings desizing and desiccation of the strands are carried out prior to testing so as to have bare, dry strands. A given length L (m) is then cut off and weighed to obtain M (g). Standards: ISO 1889, NFB 38-103 or DIN 53830.

The average linear mass based on several specimens is:

$$m = \frac{M}{L} \quad (g/m).$$

The results from ISO 1144 and NFB 01-001 standards are usually expressed in tex. Tex is the mass in grams of 1000 metres of desized and desiccated strand. The Tex, T, of the strand is then:

$$T = \frac{1000\,M}{L}$$

where M is in milligrams, and L is in millimetres. The number of strands in the warp or weft directions can also be counted. Standards: ISO 4602 or 7211/2, DIN 53-853, ASTM D 3775 or NFG 07-155.

2.1.4 Thickness check

This is carried out with callipers or a ruler; tolerance is $\pm 15\%$ of nominal thickness (mm). Standards: ISO 5084, NFG 07-153, DIN 53855, ASTM D 578, ISO 4603 or NFG 07-104.

2.1.5 Coupling size chemistry

Size chemistry involves all the products applied to the yarn chiefly to obtain a satisfactory physical/chemical bond with the resin matrix. Specific chemical sizes are available for each resin: e.g. polyester compatible sizes, epoxy compatible sizes.

(a) Size type

The nature of the size is of the highest importance since if correct it will ensure good shear strength and resistance to moisture/water. Glass fibre is usually sized with either a polyester or epoxy compatible coating. Polyester size is only compatible with polyester resin. Epoxy size is especially recommended for use with epoxy resin although it is also compatible with polyester resin. Carbon fibre is generally surface treated by hot air or chemical oxidation and then given a protective/bond enhancing coating. Lastly, sized fabrics that have been subjected to 'textile' sizing must be desized to ensure good bonding to the resin's matrix.

The greatest care must therefore be taken to use only a type of size compatible with the resin used for moving the laminate. This is very important. Although the sizing film is very thin ($<\mu m$), it can be analysed chemically (e.g. by infrared spectrometry) to determine its nature. The type of size should always be checked against the manufacturer's specification on delivery.

(b) Size ratio

In the case of glass fabrics and rovings, the quantity of organic or combustible size is given by the following standards: NFB 38-101, ISO 1887, DIN 52.330, ASTM D 678. The test process consists in burning off the size film and weighing the material before and after calcination. The test is carried out on five samples of glass rovings or fabric placed on a heat-resistant stand.

First the support stands are placed in an oven at 650 °C. After cooling in a desiccator, the stands are weighed (mass m_0). The samples are placed on the stands, which are then put into an oven at 105 °C for one hour in order to eliminate all traces of humidity. After being cooled in a desiccator, the samples and stands are weighed together: mass m_1.

Then the samples on their stands are burned off for five minutes at 650 °C in an oven, the door of which must remain open. Afterwards, with the oven door closed, they are left in the oven for an additional 30 minutes. After being cooled in a desiccator, the samples and stands are weighed: mass m_2.

The size ratio, t, is expressed as:

$$t = \frac{m_1 - m_2}{m_1 - m_0} \quad \text{(in \%)}.$$

Carbon yarn and carbon fabric

AFNOR NFT 25-104 and ISO CD 10548, which are preliminary standards, will provide the figure for the quantity of size. The process consists in dissolving the film or size in a solvent and comparing the mass before and after dissolution. For the measurement to be accurate, it is necessary either to perform the tests on oven-dried fibres, or to measure the humidity content of fibres before the test.

If the tests are carried out on oven-dried fibres, proceed as follows. First the container must be dried for two hours at 105 °C in an oven, then weighed, m_1.

A sample of sized moist fibres is then placed in it. The sample plus container is dried for two hours at 105 °C and weighed: m_2, giving the dried sample weight, $m_2 - m_1$.

The sample is desized by soaking it for at least two hours in concentrated sulphuric acid, and by washing it afterwards for 15 minutes in boiling water and for 12 hours in cold water.

The desized sample is then placed in an oven-dried container, mass m_3, and dried for four hours in an oven at 105 °C in order to eliminate any remaining traces of moisture. The container and sample are weighed after being cooled in a desiccator, mass m_4.

The size ratio t, is:

$$t = \frac{(m_2 - m_1) - (m_4 - m_3)}{m_2 - m_1} \quad \text{(in \%)}.$$

Note that it is also possible to wash the fibre in a boiling solvent such as dimethylformamide (DMF) and a ketone. Whether DMF or H_2SO_4 is used, it is essential that all bits of fibre be recovered and weighed accurately.

Aramid fabrics and fibres
Burning the sample is impossible as it would destroy the material. However, it is possible to use cold sulphuric acid (H_2SO_4) as for carbon fibre.

2.1.6 Moisture content

The fibre matrix bond depends on the size ratio, but is impaired by moisture. In order to avoid any problems, it is better systematically to dry all fabrics in an oven before impregnating them with resin. Drying is carried out for two hours at 105 °C in a ventilated oven.

2.1.7 Density of fibre

Measuring the percentage of air bubbles in a laminate requires that the volumetric mass of both resin and fibre should be known. The measuring of the density of glass fibre, carbon fibre or aramid in the form of a continuous strand is subject to AFNOR standard T25-100 and ISO 10119. When checking, the sample will be weighed when immersed in a liquid.

The method consists of weighing the material in air, and then immersed in dichloroethane. This possesses good wetting power and a density close to that of the fibre being used.

The fibres, in 400 mg coils, are conditioned for four hours at 23 °C, 50% RH. They are then placed in a scale pan; mass in air is m_1. The weighing process falls into three stages:

1. coil and pan in the air: mass m_3 (Figure 2.1);
2. pan only, immersed in dichloroethane: mass m_2 (Figure 2.2);
3. coil and pan immersed in dichloroethane: mass m_4 (Figure 2.3).

Note that it is essential that all air bubbles should be removed from immersed mountings and coils.

The density d of the dichloroethane having been determined in advance, the density of the fibre is expressed by the following equation:

$$\rho = \frac{(m_3 - m_1) \times d}{(m_3 - m_1) - (m_4 - m_2)} \quad \text{(g/cm}^3\text{)}.$$

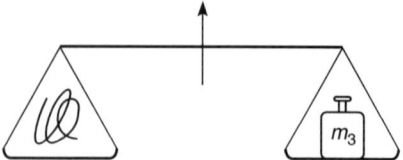

Figure 2.1

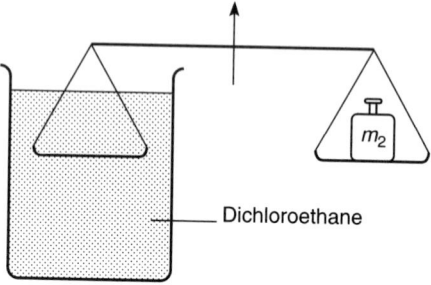

Figure 2.2

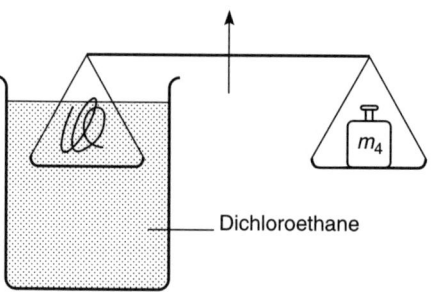

Figure 2.3

2.1.8 Mechanical resistance of rovings and strands

Strands or rovings are used to make filament wound articles. In order to predict the strength of the artefacts obtained by filament winding, one may, to a first approximation, assume that strength is due to the rovings, the additional strength due to the resin matrix being negligible. Roving strength will thus be measured by carrying out the following standard trial: ISO 3341, NFB 38-110, DIN 53834 T1, part 1. Ten samples will be used for the test.

The rovings will first be conditioned for four hours at 23 °C, 50% RH in compliance with the following standards: ISO 139, NFG 00-003, DIN 50014 and 53820, BS 1051. Each sample will have a minimum length of 600 mm.

Once mounted in the clips, a length of the roving $L_0 = 500$ mm is marked off. The cross-head movement is at the rate of 50 mm/min and the test will continue until failure of the yarn occurs.

The following information is required:

- Fr: break load (in N or N/tex);
- L (mm): the length between the marks when the break occurs;
- the average result based on the test carried out on ten rovings will be expressed as load (Fr) at a strain (A) when break occurs, where $A = (L - L_0)/L_0$ (in %).

Note that the results of this tensile test are inaccurate as the rovings, being made up of yarns of unequal lengths, and which are not all parallel, do not rupture all at the same time.

The tensile stress–strain curve is therefore not perfectly linear but there is no reason why it should be, since strength is **not** linear to failure. Moreover, the clamps tend to cut into the fibres and the fibres may slip out. Only the breaks occurring at more than 10 mm from the clips will therefore be taken as indicating a true failure.

A better method to avoid slippage is to embed the ends of the fibres in a resin.

2.1.9 Mechanical resistance of fabrics

Suppliers guarantee the strength of their fabrics, the result usually being quoted per unit of width. This quality can be determined by carrying out a tensile test. The result obtained may be used to predict the strength of laminated materials on the basis of the fibre content, ignoring the contribution of the resin.

It is useful to measure both the warp and weft strengths according to the following standards: ISO 4606, NFB 38-203 or DIN 53857. The test involves five weft samples and five warp samples. Strips of fabric 50 mm wide and 350 mm long are cut and the extremities bonded on to pieces of thick, rigid paper, which can be secured in the grips. The samples should be conditioned for four hours at 23 °C, 50% RH. The tensile test is performed at a speed of 100 mm/min until failure occurs. The breaking load Fr is measured (in N).

The final result is expressed by relating the load Fr to the width of the sample: Fr (in N) for a piece of fabric 50 mm wide. Only failures occurring more than 10 mm from the clamps are acceptable.

- The clips must have no teeth and must be covered with rubber.
- Very low results involving fibre pull-out or failure close to the grips should be discarded.
- The results obtained may be inaccurate and should be regarded as only indicative of the strength.

2.2 RESIN CONTROLS

Resins are organic products made by a generally trustworthy heavy chemical industry. However, some resin suppliers formulate systems by mixing different components from various sources. Small variations in the formulation either of the resin or the mix used for each product batch cannot easily be detected but can entail significant variations in the behaviour of the laminate.

The central laboratory of Aérospatiale (France) has demonstrated that a very small variation in the formulation of epoxy resin can lead to a 30% reduction in the life expectancy of a laminate. Cured resins and the uncured components are subject to ageing. The physical and chemical make-up changes and raw materials are very susceptible to storage conditions: duration, light exposure, moisture, temperature, etc. Clearly these effects can be minimized by storing materials in airtight containers.

Ideally, each time resin is supplied, it should be thoroughly analysed, and any variations from the original specification should be noted. This is easier when the supplier provides the original details of the resin. This type of analysis can be carried out using IR spectrophotometry, NMR (nuclear magnetic resonance) or gel chromatography (GCP).

Unfortunately, suppliers may not be willing to give full details of resin formulation, molecular weight distribution, etc., for comparison purposes. Furthermore these analytical methods are very costly and only very large plants have the capacity to execute them. Small companies, on which the future of composite materials is based, cannot generally afford these sophisticated techniques. In such cases cheaper but valuable techniques based on viscometry or differential scanning calorimetery (DSC) may be used.

Here we shall limit ourselves to tests performed on unsaturated polyester resins (UP) and on epoxy resins (EP), which account for over 95% of the matrices used in the composites market.

2.2.1 Chemical controls

(a) IR spectrophotometry

The molecules of the polymer have characteristic vibration frequencies v, which can be excited by infrared radiation of wavelength: λ, 10^{-2}–10^{-4} cm. In terms of the wavenumber, $1/\lambda = v$, the values range from 100–10 000 cm^{-1}. Ordinarily, spectrographs used for analysing organic matter work in the 500–5000 cm^{-1} range.

The resin to be analysed is spread in a thin layer on a chip of potassium bromide, KBr. The IR beam will be partially absorbed by the resin layer. Absorption in the KBr chip is determined in a null experiment.

The detector, illustrated in Figure 2.4, receives beam (1) of intensity I, which has been partially absorbed, and compares it with beam (2) of intensity I_0,

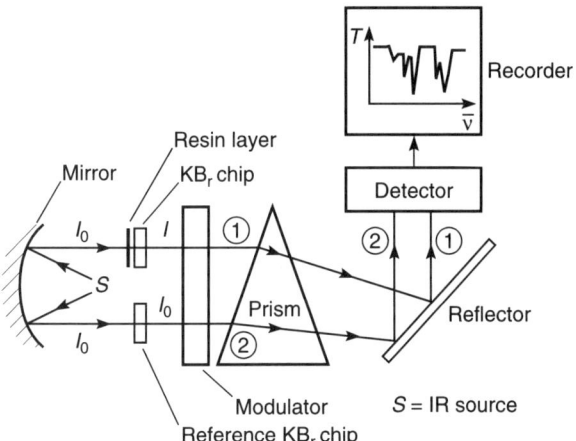

Figure 2.4 IR spectrograph.

which has interacted with the resin-free KBr chip. The result may be expressed either in terms of the absorption ratio or the transmittance ratio.

The transmittance ratio is $T = \dfrac{I}{I_0}$;

The absorption ratio is $A = \log \dfrac{1}{T} = \log \dfrac{I_0}{I}$.

There exist good computer data banks which give the function of $\bar{v}$ relative to transmittance (T) or absorption (A).

Each value of $\bar{v}$ is characteristic of a particular type of molecule, e.g. absorption at $\bar{v} = 1250 \, \text{cm}^{-1}$ indicative of the presence of epoxy groups. Note: there may be several values of $\bar{v}$ for a particular molecular species.

Identifying the type of resin
The computer linked to the spectrograph will compare the spectrum of the sample with those in the data bank and will indicate the most probable type of resin. Figure 2.5 illustrates the results for:

- an unsaturated polyester resin (see Figure 2.5(a));
- an epoxy resin (see Figure 2.5(b)).

Standard: ANFOR NF T51-500, NF T51-528 and NF T51-529.

Quality control
Whenever resin is delivered, it is advisable to make sure that the chemical composition of the batch is identical to that of the reference material. The

spectrum obtained must be exactly identical, with regard to $\bar{v}$, to that of the original batch.

The spectrum can give evidence of the deterioration of the quality of the resin due to water absorption, solvent evaporation, etc. The results for fresh resin are given in Figure 2.5(c); those for six-month old resin are shown in Figure 2.5(d).

Gel permeation chromatography (GPC)

This technique evaluates the molecular weight distribution in the sample. The resin is diluted in a solvent and is then injected into a set of columns

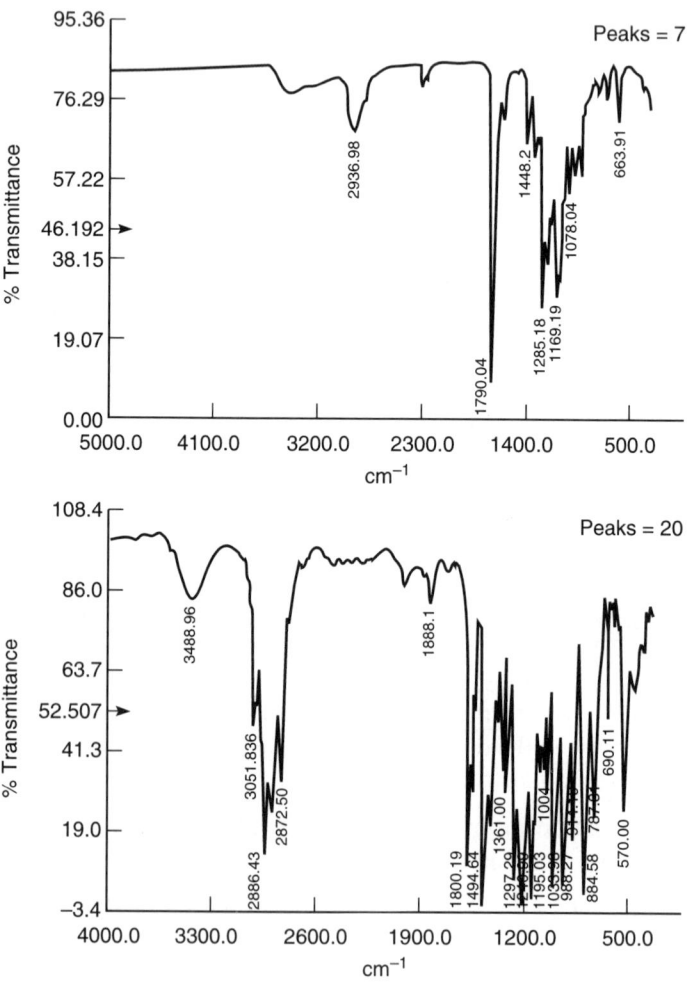

Figure 2.5 IR spectro-diagrams: **(a)** fresh resin transmittance; **(b)** fresh epoxy resin transmittance.

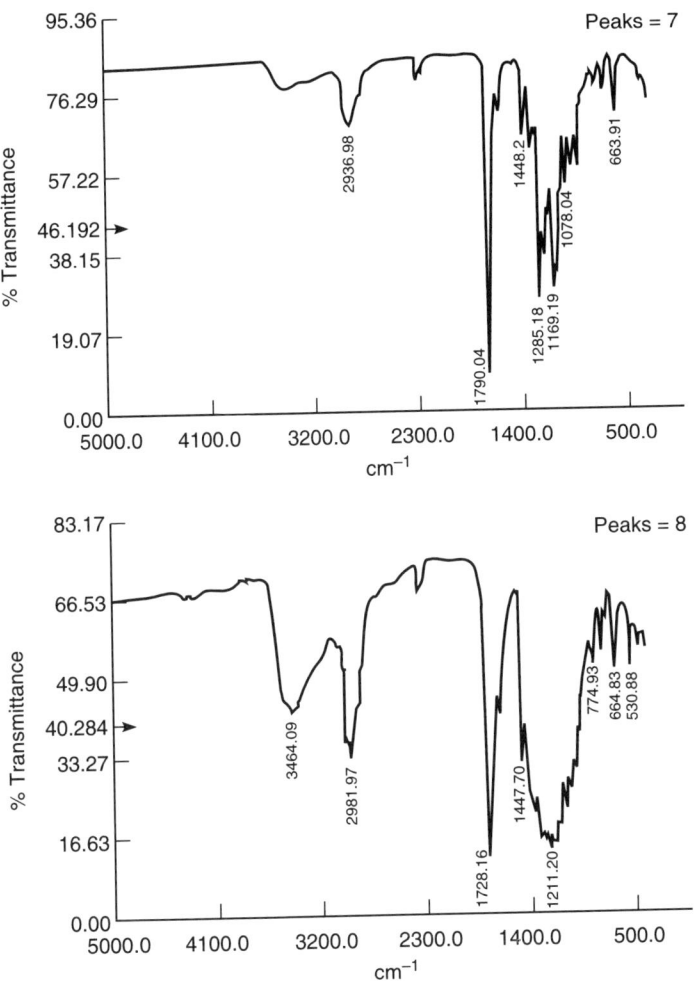

Figure 2.5 (*cont'd*) (**c**) fresh resin transmittance; (**d**) transmittance after six months' ageing.

each containing a gel of known porosity. Porosity varies from one column to another so that if the columns have been well chosen, it will be possible to separate the molecules according to their length (or mass). The result is a distribution of the number N or percentage of moles relative to the molar mass. This is given as a series of overlapping Gaussian curves as shown in Figure 2.6.

A comparison between the chromatographic spectrum obtained from the analysis of the different samples of resins makes it possible to check whether

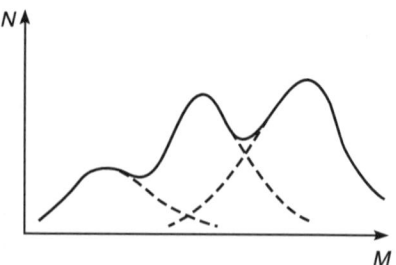

Figure 2.6 Gel permeation chromatography.

quality remains the same and is in accordance with the original standard molar distribution. The two methods used in conjunction provide an excellent way of ensuring resin quality. Standards: AFNOR NF T51-505 and NF T51-506.

2.2.2 Controls on unsaturated polyester resins

(a) List of recommended tests

Determining:	AFNOR	ISO
Density of non-thixotropic liquid resin by pycnometer method	NFT 51-201	1675
Viscosity at room temperature (Brookfield)	NFT 51-210	2555
Viscosity with rotational viscometer	NFT 51-211	3219
Content of volatile matter and styrene	NFT 51-515	7078
Conventional acid index	NFT 51-511	2114
Gel time at 25 °C	NFT 51-512	2535
Conventional reactivity with benzoyl peroxide at 80 °C	NFT 51-514	584
Conventional reactivity at 130 °C	NFT 51-518	
Total chlorine content		4615
Content of thixotropic filling – determined by loss on burning	NFT 57-557	
Temperature of deflection under load	NFT 51-005	75
Volume shrinkage	NFT 51-501	3521
Charpy impact resistance	NFT 51-035	179
Flexural characteristics, i.e. E modulus and break strength σ	NFT 51-001	178
Hydroxyl index	NFT 51-513	2554
Linear shrinking	NFT 51-401	2575

For further information and procedural details see reference standards.

In order to keep the number of tests to the minimum, we shall use as a reference the *Technical specification for unsaturated polyester resins* used by the French Ministry of Defence which recommends that the following tests at least be carried out whenever a resin shipment is delivered:

- determination of acid index;
- determination of styrene content;
- gel time at room temperature;
- determination of viscosity at 20 °C;
- determination of the amount of thixotropic filling.

The department can request that the supplier carry out new tests when storage exceeds six months (at a temperature below 22 °C) in insulated tanks, with a view to verifying that the resin has not deteriorated.

(b) Determination of acid index

This determination is based on NFT 51-511 and ISO 2114 or DIN 53-402, BS 2782 part 4 method 432B. The acid value is the number of milligrams of KOH necessary to neutralize one gram of unsaturated polyester (UP) resin. The procedure consists of adding a known weight of resin, m, to 50 ml of solvent (a mixture of toluene, ethanol and acetone), which will dissolve the resin when heated. A few drops of an indicator, thymol blue, is added and the solution titrated with 0.1M KOH, volume V_1. The same operation is then repeated with 50 ml of resin-free solvent, requiring a volume V_2 of KOH.

The acid value of the UP resin is:

$$I_a = 56.1 \times \frac{(V_1 - V_2)}{m} \times 0.1.$$

For further details see standards.

Example

For a UP resin one gets: $I_a \leqslant 20$ mg of KOH/g resin. An excessive acid index is indicative that not all the acids or anhydrides reacted with the polyols when the resin was manufactured – there is too much acid. By getting between the macromolecules of polyester during network formation, these molecules of acid disrupted the linking of prepolymer molecules. The network is consequently incomplete, resulting in a lower strength and modulus than anticipated.

To determine the type of acid IR spectrophotometry can be used. The polyester resin (UP) is spread into a thin layer on a KBr or NaCl slide and the sample heated at 40 °C under vacuum causing the styrene to evaporate. The slide is then placed in the IR spectrophotometer. The IR output shows the characteristic peaks of acids used in manufacturing the resin:

- maleic anhydride;
- isophtalic anhydride; etc.

Note: the hydroxyl index can also be determined by using the following standards: ISO 2554, NFT 51-513, DIN 53-240 and 16945, BS 2782 part 4 method 432B.

(c) Control of styrene and volatile matter content

This method determines the weight ratio of styrene contained in polyester resins which are supplied in a styrene solution. Controls can be carried out weekly.

The principle is that all the styrene contained in the UP resin is evaporated and the weight loss is then measured. Resin is placed in a Petri dish, giving a total mass m_1. Note this must be accurately weighted to within 1 g. Acetone is then added (about 10 ml) to dissolve the resin and allow the styrene to evaporate. The Petri dish is then placed in a ventilated oven at 125 °C for 2 hours 30 minutes until the acetone and styrene have completely evaporated. Afterwards, the dish and contents are dried and weighed accurately, giving a mass m_2.

Three replicate tests are used to determine the weight ratio of styrene:

$$S = \frac{m_1 - m_2}{m_2} \ (\%).$$

For a UP resin one gets: $S \leqslant 45\%$. If this value is markedly lower or decreases with time it is a clear indication of styrene evaporation. As styrene is the linking agent of prepolymer polyester molecules, a lack of styrene will lead to a lesser degree of cross-linking. The cured resin will therefore have a lower strength and modulus.

Reference standards: NFT 51-515, DIN 16945, ISO 7028.

(d) Control of non-volatile filling

The aim of the test is to determine the silica (SiO_2) content of polyester resins (UP). The calcination of the resin, together with the evaporation of all volatile matter leaves the silica filling intact.

A heat-resistant and dry crucible (mass m_1) is weighed accurately. UP resin mass $m_2 \simeq 10$ g is poured into the crucible. The crucible is placed in an oven at 625 °C for 30 minutes to burn off the resin and evaporate the volatile matter. After heating, the crucible must be kept in the oven at 625 °C until the mass is consistent. In general, 15 minutes is necessary to eliminate all traces of soot.

The crucible is then cooled in a desiccator and weighed accurately, giving a mass m_3, so that m_4, the mass of silica, is given by

$$m_4 = m_3 - m_1.$$

The weight ratio of silica based on these tests will therefore be:

$$SiO_2 = \frac{m_4}{m_2} \; (\%).$$

Resins usually contain 1–2% of thixotropic filling matter.

Standard reference: NFT 57-557 and NFT 57-571: loss through calcination.

Standards known as 'loss through calcination standards' apply to pre-impregnated fabrics and glass fibre roving. The method recommended by AFNOR is similar to the one used above for resin filled with SiO_2.

(e) Gel time at room temperature

It is necessary to determine the gel time, Gt, because each laboratory tends to have its own definition.

Physico-chemical definition
Gt is the time after the initiation of cure at which a three-dimensional network starts to form throughout the mix due to the linking of prepolymer molecules, as shown in Figure 2.7. When gel takes place, resin viscosity increases and a significant stiffness becomes apparent.

Conventional definition
According to standards NFT 51-512, ISO 2535, DIN 16945 and BS 2782, Gt is the standard period at the end of which viscosity reaches 50 Pa s (pascal second) or 500 poises at 25 °C. The polyester septim is a mixture of resin + catalyst + accelerator in the usual proportions (0.1% of cobalt octoate containing 6% of Co and 1.4% of PMEC (with 9% of active O or preactivated resin + a catalyst in the case of resin which is already preactivated).

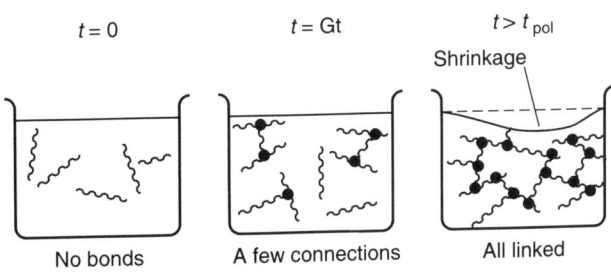

Figure 2.7 How polymerization takes place.

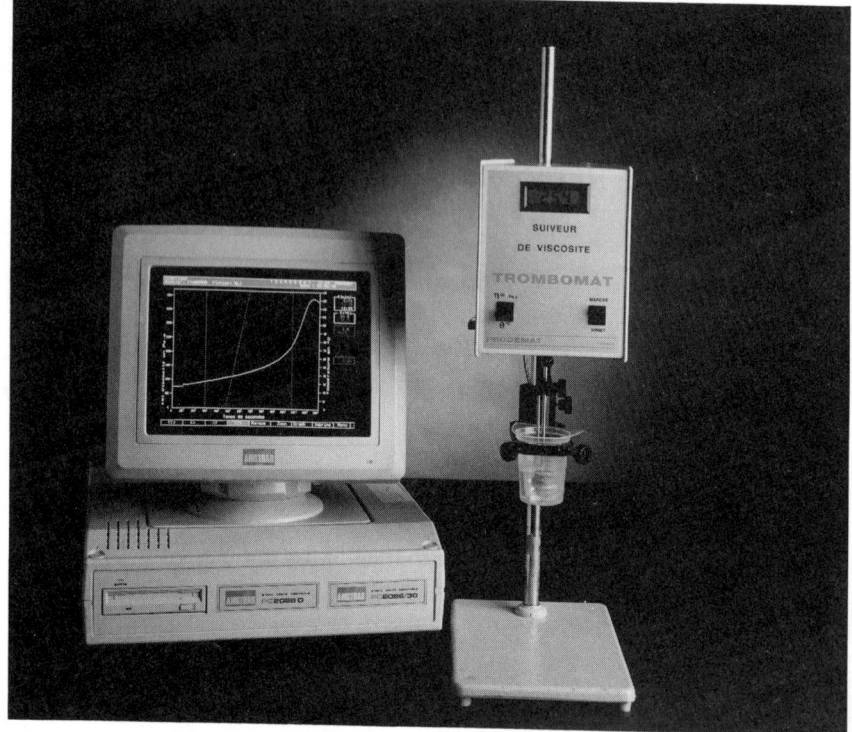

Figure 2.8 Trombomat®.

The pot life is the period of time which elapses between the time when the resin is mixed with the catalyst and the time when the mixture gels. It is known as Gt, the gel time at room temperature. In practice, it is the time the user has at his/her disposal to impregnate the fabrics, degas and press them.

Methods to determine Gt at room temperature
In the ISO standard method 50 g of pure resin is poured into a test tube which is then placed in a water bath at 25 °C. An automatic pipette is used to add 0.5 ml of accelerator and 0.7 ml of catalyst.

The viscosity of the mix is determined with a Brookfield viscometer, as illustrated in Figure 2.9, rotating at 10 revolutions per minute. The gel time at 25 °C is the elapsed time taken for the dynamic viscosity, η, to reach 50 Pa s. The change of dynamic viscosity η with time is recorded. Note: if a gel time is required under normal working conditions, the heated water bath is not required.

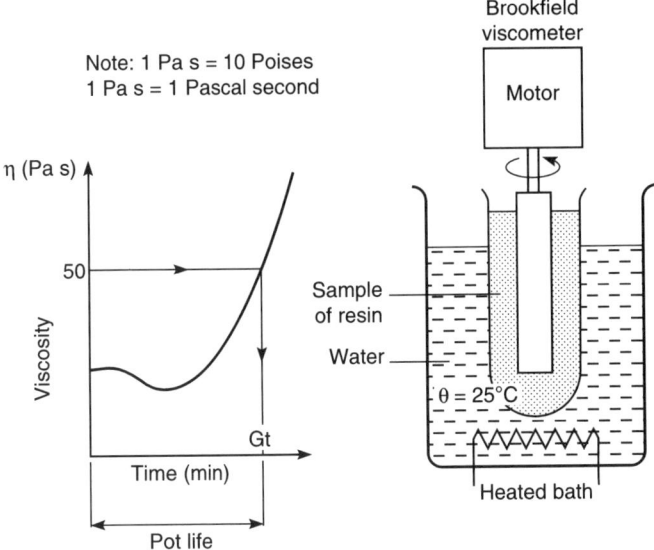

Figure 2.9 Determination of standard gel time (Gt).

The gel time, Gt, is the average of two readings. For a typical UP resin, the Gt of pure resin is $20 < Gt < 30$ minutes.

This method can be applied to any resin. It also allows the detection of resin ageing due to prereaction.

The room temperature gel time can be measured with an automatic gel detector (at 20 °C), which determines the value as a function of the percentages of catalyst and accelerator present. This works on the following principle. A predetermined mixture of resin, catalyst and accelerator is poured into a test tube which is placed in a water–oil bath at 20 °C and the viscosity of the mix monitored as a function of time with a reciprocating piston. The gel time is defined as that when the piston no longer moves.

The device has three test tubes as shown in Figure 2.10, which either allows an average result for any test or for the study of the influence of catalyst and accelerator percentages on gel time.

For example, 50 g of UP resin activated with 0.5 ml of Co octoate (0.1% of the weight) and catalysed with 0.7 ml of PMEC (1.4%) has a Gt of 30 minutes.

The automatic Trombomat® viscosity recorder is another device that can be used. It is designed to study and record the kinetics of polymerization in liquid resins. A blade moving sinusoidally and horizontally is linked to a force sensor. The blade dips into a pot containing the catalysed and

Figure 2.10

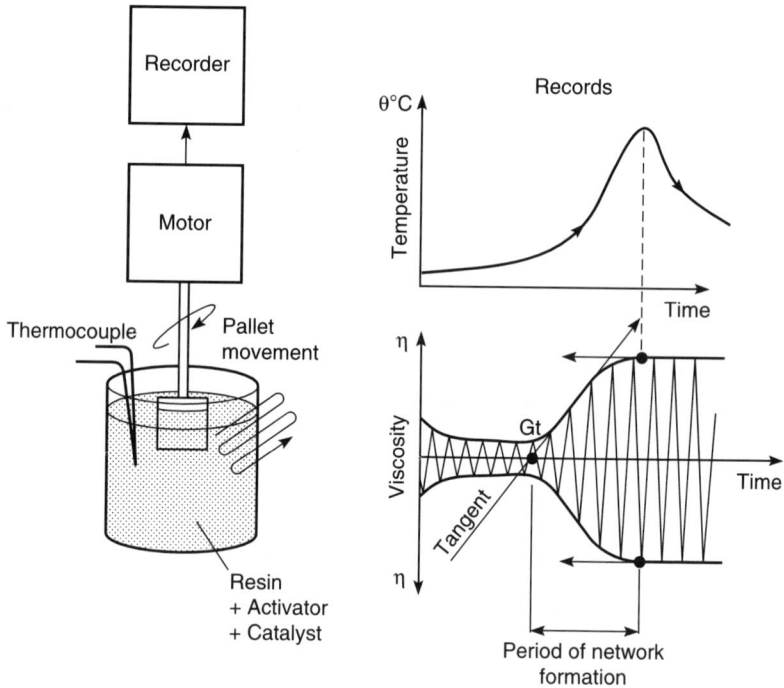

Figure 2.11 Viscosity record (from a Trombomat®).

accelerated resin. The mass of the mixture is constant. As the resin polymerizes the apparatus gives a plot of force against temperature, as shown in Figure 2.11.

The heating curve (due to the resin exotherming as it cures) allows the ageing of the resin to be checked, provided a constant mass is used. The gel time (Gt) is determined through the tangent method. The origin is the time when the resin was mixed. The blade is coated with vaseline so that it can be extracted from the block of polymerized resin.

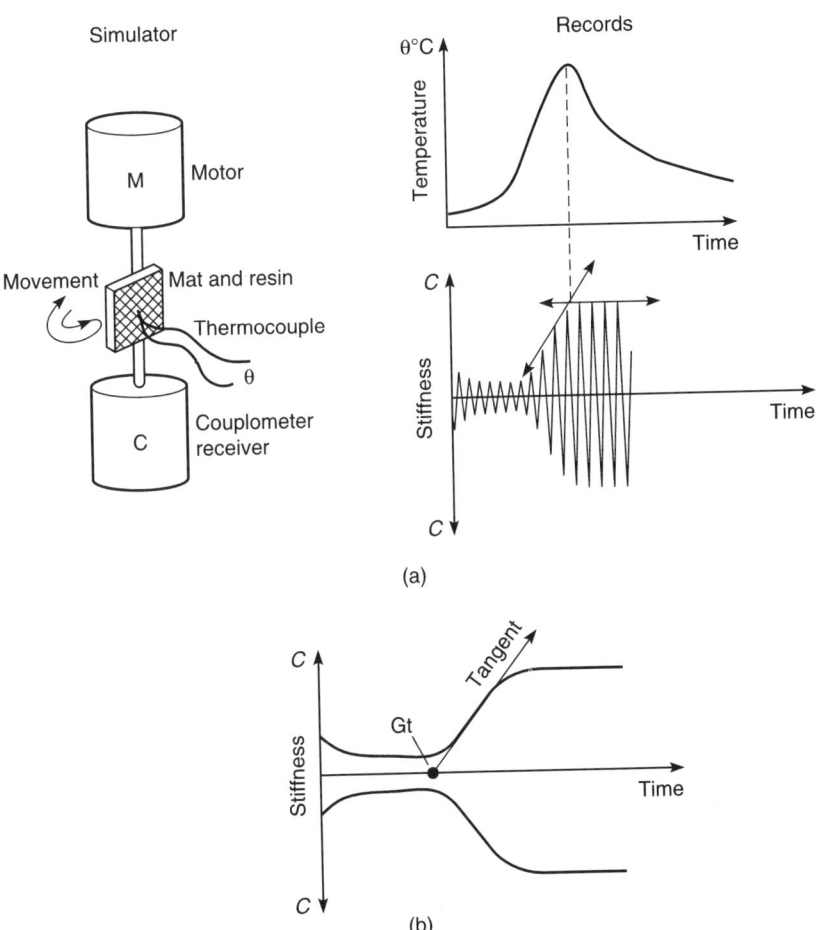

Figure 2.12 Curing simulator (Kinemat®): **(a)** experiment involving the use of a curing simulator; **(b)** determination of gel time through the tangent method.

The Kinemat® was originally designed to simulate the curing of prepregs, as illustrated in Figure 2.12. It can, however, be used to study the polymerization of any kind of thermosetting resin and to determine Gt.

A small sample of $50 \times 15\,mm$ is made by laying up three layers of glass fibre mat weighing $450\,g/m^2$ or $300\,g/m^2$. The sample is impregnated with the mixture of resin, accelerator and catalyst, and immediately set between the clamps of the curing simulator. The start time is when the resin, catalyst and accelerator are mixed.

In the Kinemat® (shown in Figure 2.13) the sample is subjected to alternate twists with a set angle of $\pm\alpha$ and about 35 degrees. The twisting torque C is monitored as a function of time with a force transducer. A thermocouple set inside the sample between the mat layers records the temperature as a function of time.

As in the case of the Trombomat®, the gel time Gt is determined by the tangent method. The advantage of this method is that it provides information on the phenomena which take place between the different layers of laminate as polymerization and network formation occur in the resin. In particular the type of sizing of the mat (or the woven fabric) used can have an impact on gel time.

From observations of the variation of twisting couple with time, it is possible to distinguish an isophtalic polyester resin from an orthophtalic one, and an aged from an unaged resin, as shown in Figures 2.14 and 2.15.

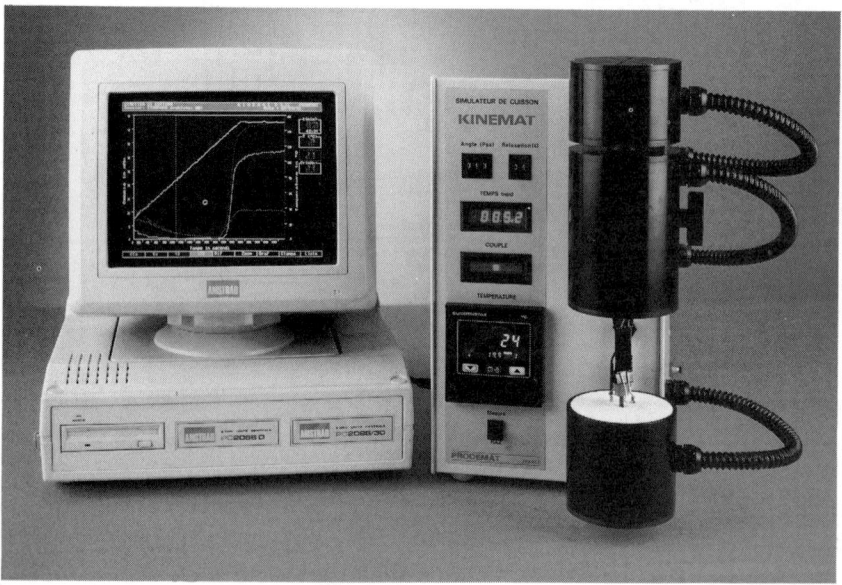

Figure 2.13 Kinemat®.

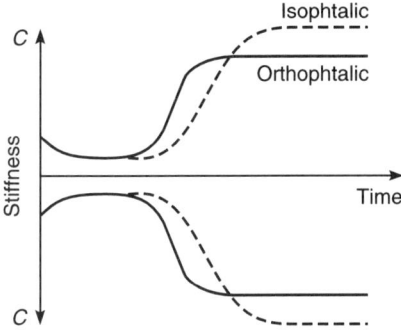

Figure 2.14 Unsaturated polyester (UP) resin spectra.

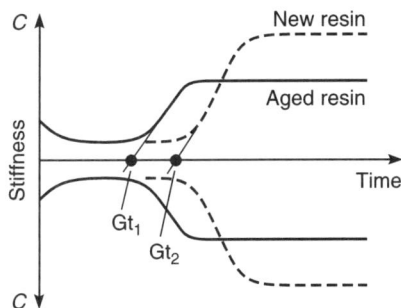

Figure 2.15 New and aged resin spectra.

An aged resin which has not lost any styrene will have a shorter gel time than a fresh resin. Moreover, the torque or stiffness transmitted will be weaker, due to a decrease in styrene bonds after a long storage time.

(f) Determination of demoulding and curing times

This is determined with a curing simulator, as shown in Figure 2.16. The same kind of sample as was used to determine Gt is employed. Curing parameters are then applied, i.e. either curing at a set temperature, θ: 80–100 °C

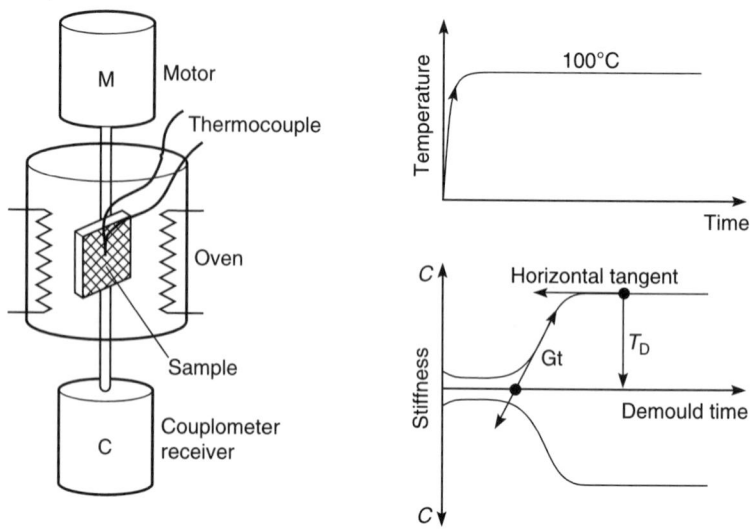

Figure 2.16 Determining the demoulding time.

for example, for a polyester resin, or 120–180 °C for an epoxy resin; or the curing program scheduled for the prepreg or industrial moulding is used.

When the Kinemat® is used, demoulding time is taken as the time at which the torque does not change further, i.e. a horizontal or nearly horizontal tangent. An aged resin has a shorter curing time because fewer molecular bonds remain to be made due to earlier chemical reactions having taken place.

(g) Testing viscosity at room temperature

Brookfield method
AFNOR NFT 51-210 or ISO 2555 standards.

This can be carried out on the resin free from accelerator or catalyst, or on preactivated resins at a predetermined temperature and humidity (e.g. 20 °C – 50% RH).

The Brookfield dynamic viscometer illustrated in Figure 2.17 (which can also be used for paints, varnish and other resins) measures the torque C_r exerted on a rotor running at a predetermined constant speed in the sample which is taken to be a non-Newtonian fluid. Note that the dynamic viscosity, η, is given by

$$\eta = 1\,\text{N}\,\text{s/m}^2 = 1\,\text{Pa}\,\text{s} = 10\,\text{poises}.$$

The SI unit is the Pa s. For an isophtalic polyester resin at 20 °C, $3.5 \leqslant \eta \leqslant 5.5$ poises at 20 °C.

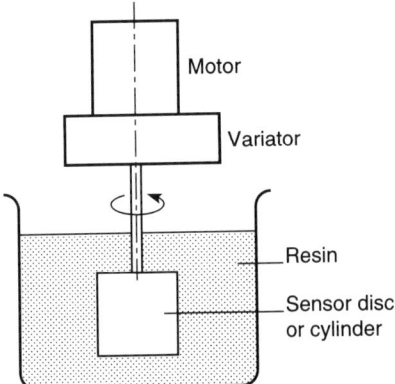

Figure 2.17 Brookfield viscometer.

Using this test it is possible to check whether a given resin:

- is thixotropic or not;
- has aged;
- contains prepolymer molecular chains.

Thus a resin filled with SiO_2 is more viscous than a resin with no silica filling. An aged resin has already started cross-linking. In a polyester resin, styrene molecules may have linked up to form small polystyrene chains, thus increasing viscosity.

Consistometric dish method
A simpler and cheaper, but also less accurate, method of checking viscosity, is to measure the time taken for a given volume, V, of resin to run through the calibrated hole in a dish under standard conditions ($20\,°C - 50\%$ RH), as shown in Figure 2.18.
The dynamical viscosity is

$$V_{(stokes)} = 0.0378\,t - \frac{4}{t}$$

where t (s) or η (poises) $= V_{stokes} \times d$ and d is the density of the liquid. Standards ISO 2431, ASTM D1200, DIN 53211, number 4.

Conclusion
Resins must be sufficiently fluid at $20\,°C$ (η between 2 and 5 poises) to be able to impregnate woven fabrics or other forms of reinforcement. The usual filler ratio is 30 to 50 w/o of glass fibre, or glass fibre plus other solids in a sheet moulding compound (CMC).

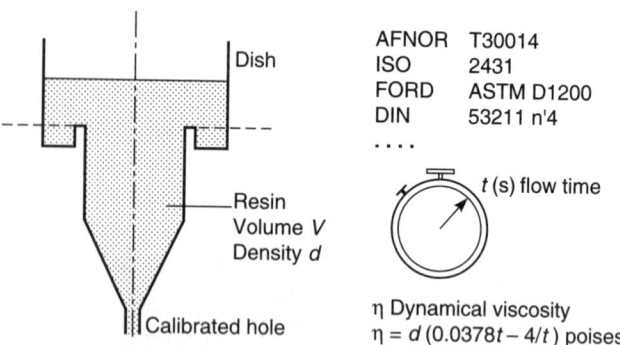

Figure 2.18 Consistometric dish method.

(h) Testing resin overall reactivity

The most suitable device is the differential scanning calorimeter (DSC), shown in Figure 2.19.

A measured mass m of catalysed and accelerated polyester resin is placed in the DSC and the exothermic heat flow in W as a function of the temperature θ reached during polymerization is recorded, as shown in Figure 2.20. The results enable the kinetics of the reaction to be determined, as in Figure 2.21.

The area A under the peak is proportional to the energy released by the cross-linking reaction, in J/g. Two different batches of resin can be compared in order to check whether their formulation has changed or whether one has aged.

Standard reactivity at 80 °C
Standards: NFT 51-514, ISO 584 or DIN 16945.

The basic principle consists of heating 100 g of polyester resin catalysed with 1% of BOP, benzoyl peroxide, in a glass test tube, and recording the change in temperature as a function of time, as shown in Figure 2.22.

The following are measured:

• the maximum temperature of the exotherm peak;
• the difference Δt_{65}^{90} from 65 to 90 °C.
• the difference Δt_{65}^{max}, also known as the gel-peak difference or Δt_{gel}^{peak}.

For example, for a fresh isophthalic UP resin one gets:

$$\Delta t_{65}^{90} \geqslant 7 \text{ minutes}; \quad \theta_{max} < 245 \,°C; \quad \Delta t_{65}^{max} \simeq 2 \text{ minutes}.$$

Figure 2.19

This test makes it possible to check resin on delivered as well as stored material. The same conditions of course must always be applied as far as mass, temperature and catalyst are concerned. For further details concerning details of the catalyst, refer to the standard.

Note: this test can be adapted for room temperature cross-linkage with PMEC as a catalyst and cobalt octate as an accelerator although the test is not a standard one when carried out in this way.

(i) Density of liquid resins

NFT 51-201 or ISO 1675.

The testing of pure resin at 20 °C is carried out with a pycnometer. For an isophthalic UP resin, the density $d \leqslant 1.12$. This test is not very useful and does not indicate if the resin has aged or has altered during storage.

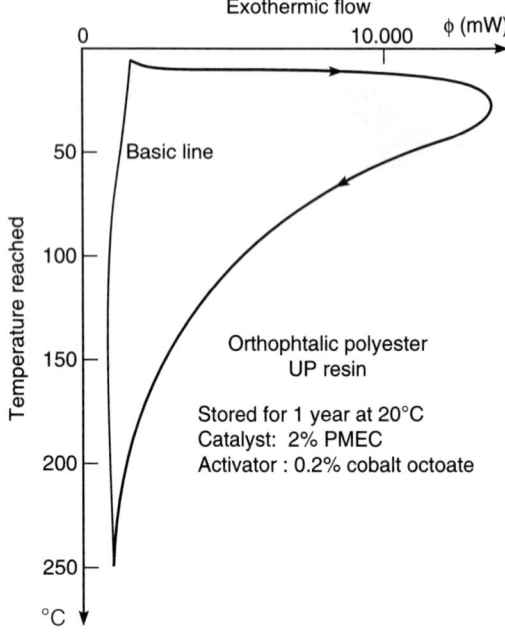

Weight	33.21 mG
ΔH Exoth	5779 mJ
ΔH	174 J/G
Peak time	28.33 min

Exothermic flow

0   10.000 ϕ (mW)

Basic line

Orthophtalic polyester
UP resin

Stored for 1 year at 20°C
Catalyst: 2% PMEC
Activator : 0.2% cobalt octoate

Temperature reached

50, 100, 150, 200, 250 °C

Figure 2.20 DSC reactivity test.

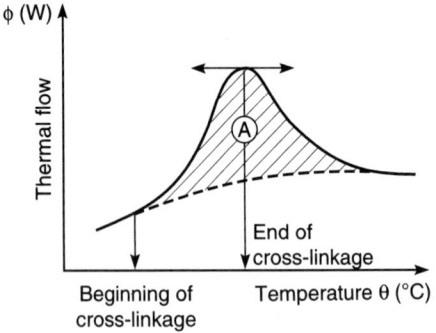

ϕ (W)

Thermal flow

Ⓐ

End of
cross-linkage

Beginning of
cross-linkage

Temperature θ (°C)

Figure 2.21 DSC reactivity.

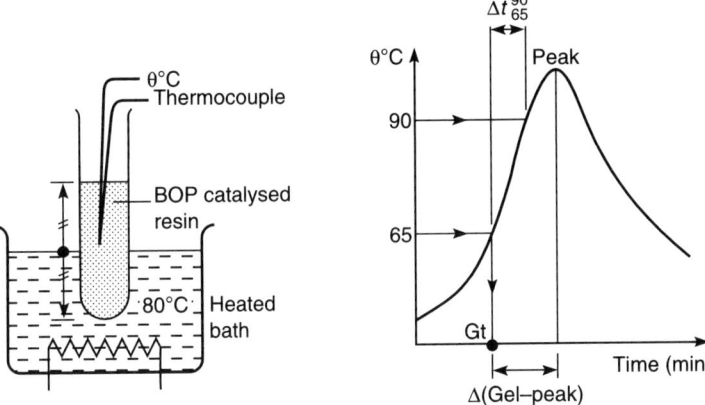

Figure 2.22 Standard reactivity.

(j) Shrinkage

The following standards: ISO 3521, DIN 16945, BS 2782 part 6, method 644 A and NFT 51-501 specify the procedure which amounts to measuring density variations before and after cross-linking.

Let ρ_0 be the specific gravity of the mix (resin + accelerator + catalyst) at 20 °C.

ρ_0 can be obtained from the following equation:

$$\rho_0 = \frac{(m_R + m_C) \times \rho_R \times \rho_C}{(m_R \times \rho_C) + (m_C \times \rho_R)}$$

where m_R is the weight of the resin, m_C the weight of the catalyst (the activator will be disregarded), ρ_R the specific gravity of the resin measured with a pycnometer at room temperature and ρ_C the specific gravity of the catalyst measured with a pycnometer. Let ρ_1 be the specific gravity of a specimen of the mix (resin + catalyst + accelerator) after cross-linking and cooling down to room temperature. The overall shrinkage is:

$$R = \frac{\rho_1 - \rho_0}{\rho_1} \quad (\%).$$

For an isophthalic polyester resin the volumetric shrinkage $R < 9.5\%$.

Mechanical testing on cured resin

Unlike tests carried out on liquid resin, which do not take too much time, mechanical tests on polymerized resin specimens require the production of fully aired artefacts. This necessitates, for a polyester:

- glass or silicone elastomer moulds;
- preparation of the mix (resin + accelerator + catalyst) using the catalyst recommended by ISO 2535, i.e. 1.4% of methyl ethyl ketone peroxide (MEKP) 50%, and 0.6% of cobalt octoate 1%;
- curing the specimen at room temperature (20 °C) for 24 hours;
- postcuring for 12 hours at 80 °C; and
- acclimatizing the specimen for 24 hours after cooling down.

A minimum of 5 specimens is needed to allow for variation in the material and to enable a standard deviation to be obtained.

Three-point bending test It is possible to measure the bending or Young's modulus E, and flexural strength σ, in this test, illustrated in Figure 2.23. The formulae given below refer to a rectangular cross-section beam, R.

$$\sigma_R = \frac{3W_R l}{2bd^2}$$

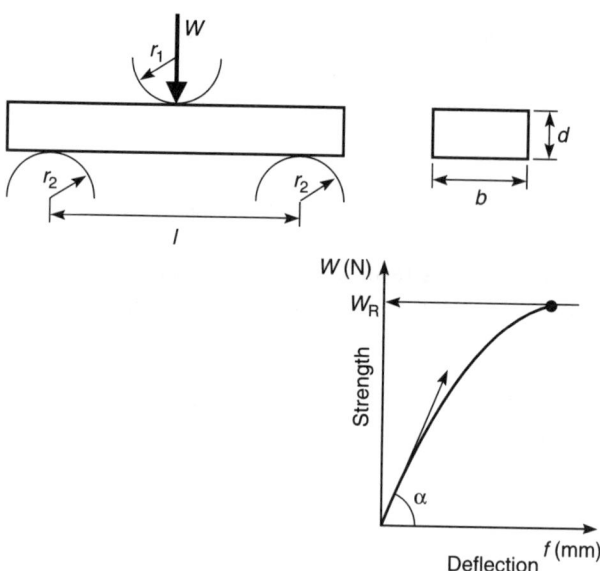

Figure 2.23 Three-point bending test.

where W_R is the load at failure, l the span, b the breadth and d the depth of the rectangular beam specimen, with W_R in N, l, b, d in mm and σ_R in MPa.

The Young's modulus is

$$E = \frac{l^3}{4bd^3} \times \tan \alpha$$

with E in MPa.

Typical values for a UP polyester resin are:

$$\sigma_R \geqslant 85 \, \text{MPa} \quad E \geqslant 3300 \, \text{MPa}.$$

Standards: ISO 178, EN 63, NF 57-105, DIN 53-452, BS 2782 part 3, method 335A and part 10, method 1005.

Charpy impact resistance test This is carried out in compliance with NFT 51-035 or ISO/R/179 with a 2 joules battering ram on a notched specimen, shown in Figure 2.24. The energy necessary to break the specimen is divided by the cross-sectional area of the specimen (0.4 cm²). Sometimes the energy is divided by the volume, V, between the supports to give an energy density. Results are frequently scattered and hence it is necessary to break a dozen specimens to get the average impact strength and standard deviation. When testing polyester resin, a reasonable value of the impact resistance or

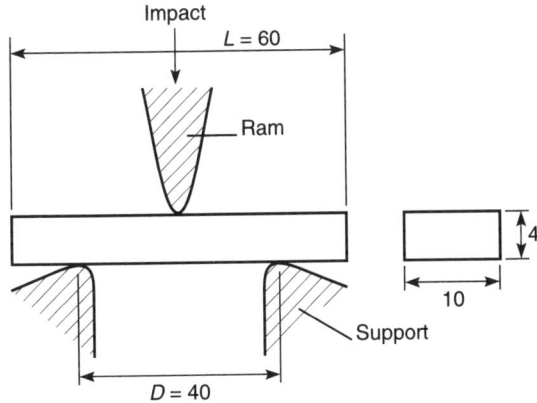

Figure 2.24 Charpy impact test.

resilience K is:

$$K \geqslant 0.4\,\text{J/cm}^2$$

Heat deflection temperature (HDT) Standards: NFT 51-005 or ISO 75.

These standards apply to resins only, not to laminates. Resin bars measuring $110 \times 4 \times 10$ mm are fully cured according to the manufacturer's schedule. The specimen, mounted on two supports 100 mm apart and loaded centrally, is immersed in a heated mineral oil bath, whose temperature rises evenly and steadily by $2\,^\circ$C per minute, as shown in Figure 2.25. The central deflection of the bar is continuously monitored. The temperature, T, at which the central deflection of a 10 mm thick bar is 0.32 mm is known as the heat deflection, or distortion, temperature. This temperature depends on the applied load W:

$$W = \frac{2\sigma b h^2}{3D} \quad \text{(N)}.$$

The flexural stress at the centre of the beam, σ, is given by

$$\sigma = \frac{3WD}{2bh^2} \quad \text{(in MPa)}$$

where $b = 4$ mm, $h = 10$ mm, $D = 100$ mm, and W is the applied load (in N).

Two cases are recognized, A and B. For the former W is such that $\sigma = 1.8$ MPa, and for the latter N is such that $\sigma = 0.45$ MPa. See Figure 2.26.

Note that this test does not enable accurate predictions to be made of the behaviour of cross-linked resins or laminates at high temperatures (see NFT

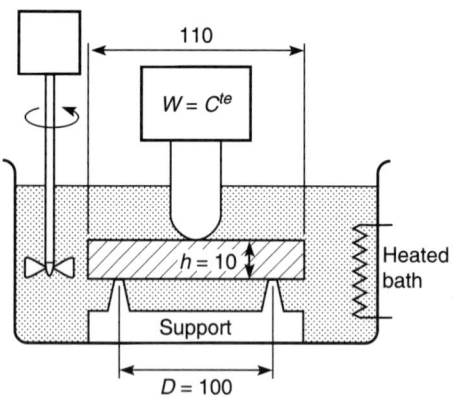

Figure 2.25 Deflection under load.

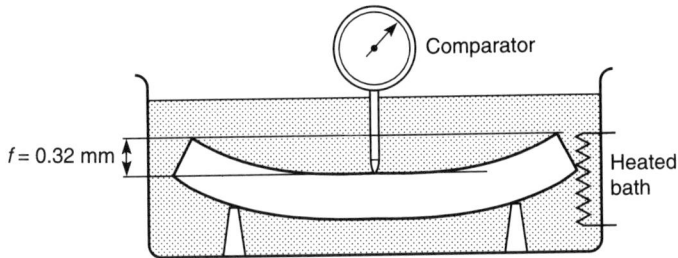

Figure 2.26 Determination of the deflection under load.

51-222). It is only a quality control test useful for comparing materials and suppliers.

For an isophthalic polyester resin cross-linked at room temperature, HDT will be of the order of $\theta \geqslant 70\,^{\circ}\mathrm{C}$.

Resin homogeneity test The homogeneity of a resin regarding its components or distribution of molecular chain lengths in the cured material can be assessed with a curing simulator. Specimens of preimpregnated fabric laminates or mats soaked in resin are cross-linked either at room or an elevated

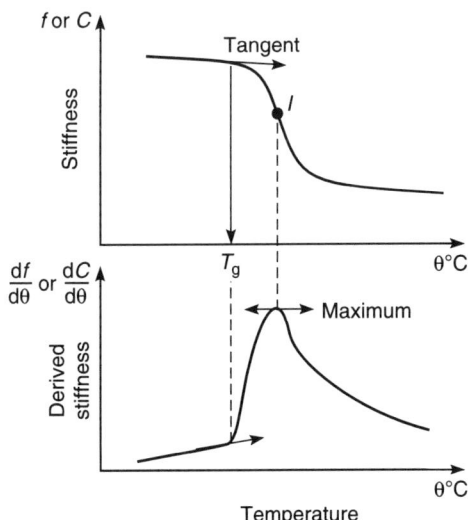

Figure 2.27 Homogeneous material.

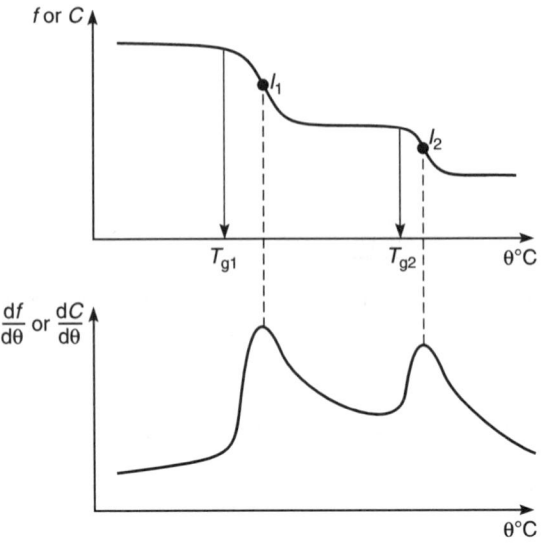

Figure 2.28 Resins which do not mix easily heterogeneously.

temperature. After being cut to the required dimensions, the specimens are placed in the simulator and subjected, depending on the type of simulator, to alternate flexure or twisting while the temperature is slowly increased. The torque C, or deflection, obtained relative to the rising temperatures θ as well as the derivative $dC/d\theta$ or $df/d\theta$ are recorded. The latter two quantities give peaks of variable width, as shown in Figure 2.27.

A drop in the torque C or deflection f that gives a loss in sharpness results from the fact that the insufficiently cross-linked fraction of the laminate has become rubbery. The drop begins at a temperature T_g, known as the glass transition temperature, T_v. A homogeneous resin is characterized by a sharp drop in C, a narrow peak $dC/d\theta$, $df/d\theta$, and a single T_g. A mixture of resins is characterized by several drops, peaks and T_g corresponding to the glass transitions of the different resins in the mix, as shown in Figure 2.28.

The peaks may overlap if the glass transition temperatures of the component resins are close. A resin whose homogeneity is poor is marked by a slow drop in f or C and a wider peak $df/d\theta$, $dC/d\theta$ as shown in Figure 2.29. Macromolecular chains albeit related are of very different sizes.

2.2.3 Control of epoxy resins

With a few exceptions, these tests are very similar to the ones carried out on unsaturated polyester resins. The curing of an epoxy resin requires the

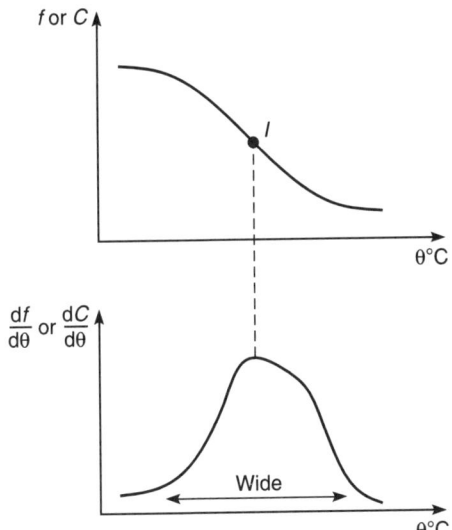

Figure 2.29 Bad homogeneity.

use of a sizeable quantity of hardener (20–40% or more by weight). This product must be tested as well. Fillers are rarely added to epoxy resins. Consequently, there is no need to measure the volatile filling content.

The aim of the tests is to verify that the resins are fit to be used:

- upon delivery;
- after being stored for a long time.

Epoxy resins and hardeners, which are principally amines or anhydrides, must be stored as follows:

- in sealed containers;
- between 18 and 25 °C;
- protected from the cold and from humidity.

If stored in such conditions, their lifetime is about one year.

Containers that have been opened must be sealed immediately after use. The repeated opening of containers is the principal cause of resin ageing and deterioration.

(a) List of recommended tests

AFNOR NFT 51-527 suggests some tests to determine the properties of epoxy resins. The characteristics to be tested fall into two categories: principal and secondary (with reference to AFNOR and ISO standards).

Principal characteristics

- Room temperature viscosity: standards NFT 51-210 or NFT 51-211, ISO 2555 and ISO 3219.
- Epoxy equivalent: NFT 51-522, ISO 3001, DIN 16945, BS 2782/4 432C and 433D.

Secondary characteristics

- Density of liquid resin: NFT 51-201, ISO 1675.
- Volatile matter content: NFT 51-525 (no ISO standard).
- Total chlorine content: ISO 46-15 (no NFT standard).
- Inorganic chlorine content: NFT 51-523, ISO 4573.
- Saponifiable chlorine content: NFT 51-524, ISO 4583.
- Tendency to crystallization: NFT 51-526.
- Reactivity (standard being studied).
- Hydroxyl index (standard being studied).

It is very important that the resin reactivity is taken as a principal characteristic if the quality control is carried out before the resin is used. Similarly, determining the gel time of the as-received resin-hardener compound is also essential if ageing after storage is to be assessed.

Controlling curing time also appears essential. Other secondary characteristics that can be measured as part of a quality control procedure are:

- shrinkage
- mechanical performance under flexure, tension and impact.

The user will obviously rely on a few methods and not employ all those listed.

(b) Controls identical to those carried out on unsaturated polyester resins

For the details see section 2.2.2.

Density of resin without any additives
The pycnometer method should be used. Standards: ISO 1676, NFT 20053, NFT 51-201.

    Examples: XB 3052 A Resin (Ciba-Geigy): $\rho = 1.16$ to 1.18 g/cm^3; EPO 661 resin (Rezolin-Hexcel): $\rho = 1.15$ g/cm^3.

Room temperature (25 °C) dynamic viscosity of liquid resin
The Brookfield viscometer method should be used with small amounts of resin without any additives. Alternatively, the coaxial rotary viscometer can be used. Standards: ISO 2555, NFT 51-210, ISO 3219, NFT 51-211.

    Example: XB 3052 A Resin: $\eta = 10$–15 poises.

Density of the liquid hardener
The same methods and standards should be used as were employed for measuring the density of liquid resin.

Example: Hardener for epoxy resin: $\rho = 0.93–0.95 \, g/cm^3$.

Viscosity of liquid hardener at room temperature
The Brookfield viscometer should be used.

Example: Hardener for epoxy resin: $\eta = 40–60 \, MPa \, s$ or $0.4–0.6$ poises.

Gel time of the resin–hardener mix
A test at room temperature can be based on the standards for polyester resin detailed in section 2.2.2. The resin and hardener must be mixed in the proportions specified by the manufacturer and poured into the test tube. The cross-linking reaction begins immediately, and the following happens, as illustrated in Figure 2.30:

- the temperature rises;
- initially viscosity decreases and then it begins to increase.

In practice, one takes advantage of this loss of viscosity in the workshop by working with lukewarm resin and prepregs. The decrease in viscosity enhances resin penetration at a relatively low temperature. Standard gel time corresponds to a viscosity of $\eta = 500$ poises (50 Pa s).

Example: EPO 661 resin (Hexcel)
Gt = 35–40 minutes.

Other tests can be carried out at warm temperatures to take into account the fact that some epoxy resins are used in warm moulds. A very useful curve, shown in Figure 2.31, is thus obtained. For example this is what is

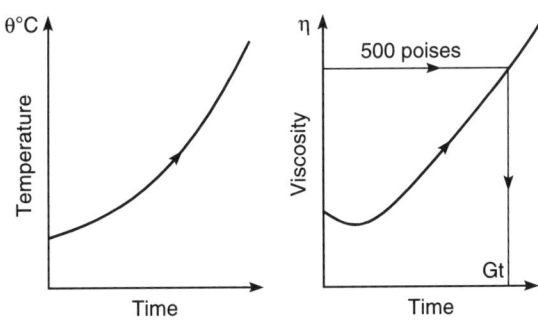

Figure 2.30 Standard gel time.

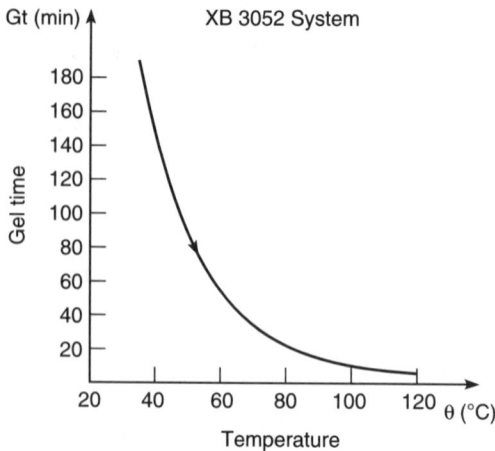

Figure 2.31

obtained for an epoxy system: we see the gel time evolution according to temperature θ.

When controlling the resin, it suffices to check a few Gt values close to the curing temperature of the laminate.

Other methods, for which there are no standards, can be used. They require the use of a viscosity analyser or a curing simulator. Gel time depends on the method used to measure it, either because the weight of resin used is not the same or because the volume of reinforcing fibres as well as their surface treatment have an influence.

That is the reason why in the field of quality control one must adopt a consistent method.

When dealing with a hot-cured (80–180 °C) epoxy resin or with hot-cured prepregs, it is advisable to determine gel time with a Vanhographe® or Kinemat® curing simulator so that real life manufacturing conditions apply. In the case of lamination by the wet method, strips of prepregs are cured in the simulator.

Pot life

The pot life of the resin–hardener mix is based on the change of viscosity in time. However, it also depends on the quantity of resin studied as this determines the heat produced by the reaction. Consequently, there is no standard to predict pot life.

For example, XB 3052 epoxy resin (Ciba-Giegy) has a pot life for 100 g of resin of 220–260 minutes and a pot life for 250 g of 65–75 minutes. Araldite 564 epoxy resin (Ciba-Geigy) for cold use has a pot life for 100 g of 30 minutes and a pot life for 250 g of 20 minutes.

Control of general reactivity of the resin–hardener mix with a DSC (differential scanning calorimeter)

For example, in epoxy resin the temperature at the beginning of the reaction $T_D = 135\,°C$; the temperature of the exothermal peak $T_p = 149\,°C$; the temperature at the end of the reaction $T_F = 173\,°C$, and the reaction enthalpy $\Delta H = 70\,cal/g$ (Figure 2.32).

These values can be modified to a considerable extent by poor storage conditions which affect the resin.

Controlling curing time

When monitoring heat-polymerized epoxy resins with a curing simulator, a cure cycle close to the one used in the workshop must be adopted. Figure 2.33 shows what happens in the case of a 130 °C epoxy resin.

The demoulding time can be inferred from this (see section 2.2.2(f)).

Shrinkage

Standards: NFT 51-501, ISO 3521, DIN 16945, BS 2782 part 6 method 644A.

Mechanical tests

These tests are expensive and time consuming. The specimens are prepared by pouring the mix (resin and hardener) into a suitable mould and curing at a temperature of 120–180 °C in an oven. The curing cycle at a particular temperature must be selected in conjunction with the manufacturer.

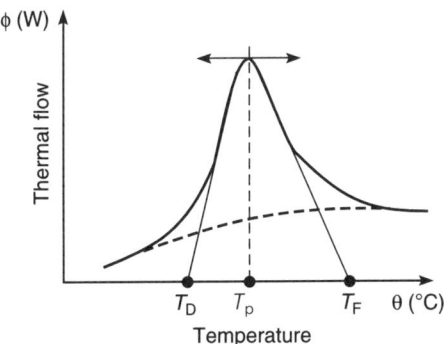

Figure 2.32 Reactivity control with DSC.

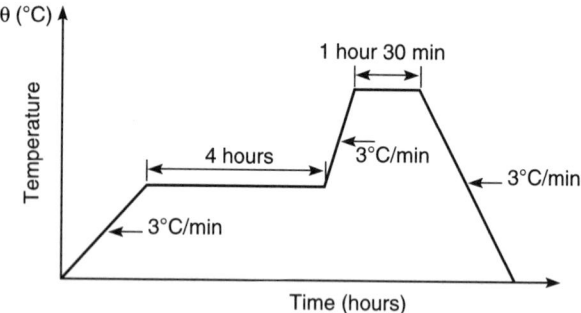

Figure 2.33 Curing cycle.

A few standardized tests are:

- three-point flexure tests according to NFT 51-001 or ISO 178;
- tensile testing according to NFT 51-034;
- Charpy impact strength according to NFT 51-035 or ISO 179;
- temperature of deflection under load: ISO 75, NFT 51-005.

For example, for a resin cured at 25 °C for one month, flexural strength, $\sigma_f = 100-110$ MPa and $E = 2500-3000$ MPa; tensile strength $\sigma_t = 75-80$ MPa; and HDT under load: 70 °C.

(c) Epoxy resins: specific controls

Determining the epoxy equivalent of the resin
Standards: ISO 3001, NFT 51-522, DIN 16945, BS 2782 part 6 method 644A.

The hydrogen bromide used reacts with the resin and is fixed by the epoxy groups. The end of the reaction can be determined by using a reagent, such as crystal violet or a potentiometer.

The epoxy equivalent, EE, is the mass of resin, expressed in grams, which contains one epoxy group molecule; alternatively, equiv/kg is the number of epoxy molecules per kg of resin.

$$\text{Epoxy group } CH_2-CH= $$
$$\diagdown\diagup$$
$$O$$

Example: epoxy resin, equiv/kg = 6.6–6.9.

Determining of amine content in hardener (no standard)

Example: Hardener for epoxy resin, equiv/kg = 9.7–9.9.

Conventional content of volatile substance
Conventional method according to AFNOR NFT 51-525 (no ISO standard)
A small mass m (g) of epoxy resin is placed in a dry dish, mass m_1 (g). After three hours at 140 °C (conventional conditions) in an oven to allow volatile substances to evaporate, the dish and its contents are cooled in a desiccator.
The new weight is m_2. A minimum of two tests must be carried out.
The volatile content, V, is:

$$V = \frac{m_1 + m - m_2}{m} \quad \text{(in \%)}.$$

For example, $V < 0.2\%$ in the epoxy resin Araldite LY556.

Thermogravimetric method A thermobalance is used to measure the variations in mass of a small sample of resin m (g) when heated up very slowly up to a point where the resin may start to decompose. The curve obtained is shown in Figure 2.34.
At T_1 °C volatile substances start evaporating; at T_2 °C the end of evaporation is reached and the loss of mass is Δm. The content of volatile substance is $V = \Delta m/m$ in %; T_3 is the decomposition temperature of the resin.

Determination of saponifiable chlorine
Standards: ISO 4583, NFT 51-524, BS 2782 part 4 method 433B.
A sample of liquid resin is used. The purpose is to determine the chlorine content from hydrochloric groups remaining in epoxy resins after manufacture. This chlorine may be present due to the original reaction not being fully completed or because insufficient neutralizing agent was used. The sample

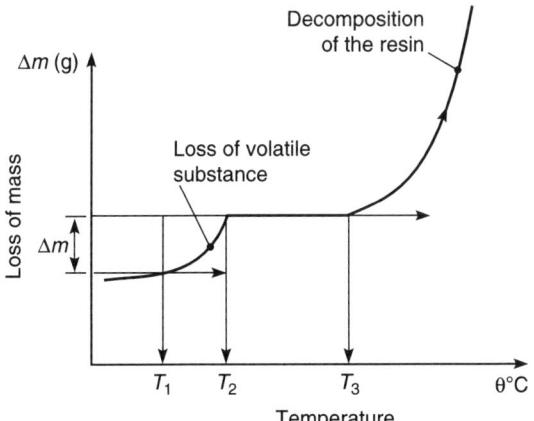

Figure 2.34 Thermogravimetric analysis.

is dissolved in hot butoxyethanol and then immersed in a cold solution of sodium hydroxide in 2-butoxyethanol for a minimum of two hours. The mixture is then acidified by an addition of butanone and acetic acid. The number of chlorine ions resulting from saponification is measured with a potentiometer using a silver nitrate solution.

For all details concerning this process, as well as the calculation of the results, see the reference standards.

Determination of inorganic chlorine
Standards: ISO 4573, NFT 51-523, BS 2782 part 4 method 433A.

The sample used is a liquid resin. The purpose is to determine the chlorine content linked to the inorganic substances contained in the resins. The sample is dissolved in cold methyl-ethyl-ketone and is then immersed in a cold solution of potassium chloride and acetic acid. The number of chlorine ions produced is measured with a potentiometer (pH meter-millivoltmeter) using an aqueous solution of silver nitrate.

For full details concerning the process, as well as the calculation of the results, see the reference standards.

Total chlorine content
Standard: ISO 4615.

2.2.4 Conclusion

There appear to be too many of these tests and they will probably be reduced by the standardization committees dealing with quality control. Each test, however, is useful for the characterization of epoxy resins.

2.3 PREPREG CONTROL

The manufacturers of reinforcing fibres tend more and more to supply rolls of preimpregnated fabrics wrapped in protective film. The resin system is B staged or partially reacted.

Preimpregnated fabrics, also known as 'prepregs', are used to produce top-quality, high-performance laminates. The resin content is low (around 30% by volume). For definitions of the terms see ISO 8604.

'Sheet moulding compounds' (SMC) are fabrics or glass mats impregnated with catalysed and heavily filled polyester resin.

Approximate composition by volume is:

- glass (usually mat): 30%;
- fillers (including MgO for maturing during storage): 30%;
- resin: 40%.

For definitions of the terms see ISO 8605.

After maturing, the viscosity of SMC is about 10^4 Pa s (10^3 poises). This value should be checked.

Prepregs and SMCs are used for low- and high-pressure moulding followed by curing. Prepregs are moulded at 7–10 bars maximum; SMCs are moulded at 70–100 bars.

Two important parameters should be verified: the amounts of reinforcing fibres, resin, fillers and solvents; and the ability to be moulded and cured. To determine these properties, standard procedures devised in most cases for glass reinforcements will be used. SMC prepregs will be dealt with more specifically in section 2.4.

2.3.1 Main standards for impregnated mats and fabrics

NFT 57-513: glass reinforced plastics (GRP). Prepreg mats of polyester. Plasticity: no ISO standard.

NFT 57-516, ISO 9780: resins and prepregs. Conventional reactivity.

NFT 57-518: GRP – prepregs glass and fillers content. Burn-off method – ISO 1172.

NFT 57-556: prepregs of GRP. Thickness of a layer after pressing and curing. No ISO standard.

NFT 57-557: glass fabric prepegs. Weight loss through calcination. No ISO standard.

NFT 57-559: prepregs based on epoxy resins. Binding time. No ISO standard.

NFT 57-570: GRP preimpregnated mats, yarns and rovings. Conventional volatile matter. No ISO standard.

NFT 57-571: GRP preimpregnated mat, yarns, roving: determination of ignition loss. No ISO standard.

NFT 57-601: GRP prepregs – mass per unit area. ISO 10352.

NFT 57-602: GRP prepregs. Conventional volatile matter. ISO 9782.

NFT 57-603: prepregs. Conventional resin flow. No ISO standard.

NFT 57-608: GRP fibre and resin content. Dissolution method. No ISO standard.

NFL 17-451: prepregs – differential scanning calorimetry.

2.3.2 Weight per unit surface area

Standards NFT 57-601 and ISO 10352. Specimens measuring $S = 100 \, \text{cm}^2$ in area are cut from the preimpregnated fabric and weighed while still in their protective polyethylene films: weight m_1(g). The protective films are then taken off and weighed accurately: weight m (g). The weight of the prepreg is $M = m_1 - m$. The weight per unit area is $M/S \, \text{g/m}^2$.

The test is carried out on five specimens.

2.3.3 Content of volatile matter

(a) After NFT 57-602 and ISO 9782

Resin impregnation often requires the use of volatile materials such as liquid solvents. These chemicals impair the quality of the cured item if there are excessive amounts of them in the prepeg material because they hamper the polymerization process and produce vapour during curing, thus creating bubbles which may cause delamination. It is therefore necessary to determine the amount of solvent remaining in the prepreg when the latter is ready to be delivered.

Specimens whose weight is M (> 10 g) are cut from the prepreg and weighed. The protective films are taken off and weighed: mass m (g). It is advisable to shred the prepreg to help the evaporation of the solvent. The weight of the prepreg is $M_1 = M - m$. The prepreg is put in an oven so as to evaporate the volatile matter:

$\theta = 80\,°C$ for two hours for an SMC polyester mat containing styrene;
$\theta_0 = 125\,°C$ for one hour for polyester prepreg;
$\theta_0 = 160\,°C$ for $\frac{1}{4}$ hour for epoxy prepreg;
$\theta_0 = 160\,°C$ for $\frac{1}{4}$ hour for phenolic prepreg.

The specimens are then cooled in a desiccator and weighed immediately after being removed. The weight is $M_2 < M_1$.

The content in volatile matter is:

$$T = \frac{M_1 - M_2}{M_1} \ (\%).$$

The test must be carried out on three specimens.

Note that the specimens may be kept in the oven until the weight is constant. When applied to polyester prepregs, this method allows the determination of the volatile styrene content.

(b) DSC method

To determine the presence of volatile matter in a prepreg, a DSC (differential scanner calorimeter) can be used. The principle of this will be described later (section 2.3.9(c)). A DSC is used to analyse a small amount of prepreg. Scanning is carried out at a low rate from a low temperature to a temperature high enough to cause complete polymerization.

During heating, the following readings should be taken:

- the fusion, endotherm, peak of the volatile matter contained in the resin;
- the vaporization, endotherm, peak of volatile solvents; and
- the polymerization, exotherm, peak for the resin.

For an example of resin behaviour, see Figure 2.35.

(c) Thermogravimetric method

Thermogravimetric analysis can be used to determine the content of volatile matter. A thermobalance is used to measure the weight variation of a small sample of prepreg subjected to a low rate change of temperature, as shown in Figure 2.36, which gives an example of the weight loss of a polyimide resin.

The following should be noted:

- at T_1 (°C), vaporization of volatile matter takes place;
- at T_2 (°C), the resin decomposes.

The weight loss due to vaporization of the solvents is: m (g).

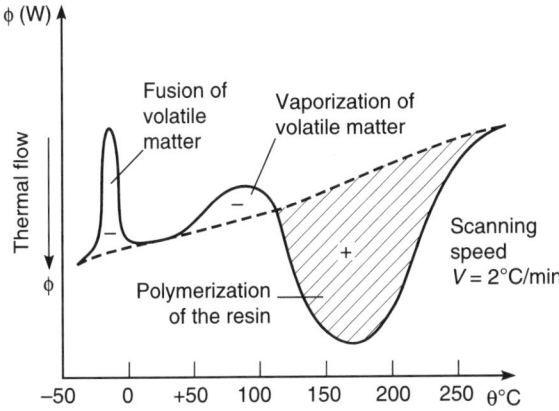

Figure 2.35 DSC method: determination of volatile matter in a prepreg.

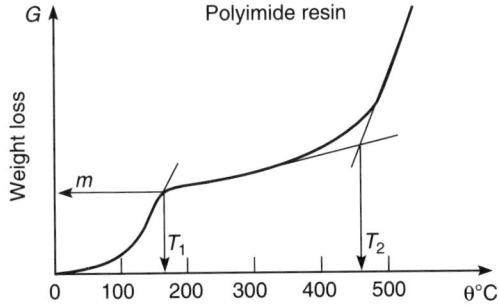

Figure 2.36 Thermogravimetric analysis (source: Aérospatiale, France).

The weight of the sample before testing was M (g). The volatile matter ratio is: $T = m/M$ (in %).

The test must be carried out on several specimens.

2.3.4 Glass fibre content

(a) Burn-off method

Standard NFT 57-557 (no ISO standard) and NFT 57-571 for prepregs (rovings, yarns, mats, etc.).

Polyester resin prepregs

This method involves taking three specimens and removing the protective films. Each specimen is weighed, M_1 (g), to give the weight of the resin and glass fibre. The specimens are then placed inside an oven to which there is a supply of air, and set alight. Burn-off is carried out at 625 °C until the weight is constant and only glass fibre remains. The glass fabric is weighed after being cooled in a desiccator, M_2 (g).

The loss due to burn-off is: $M = M_1 - M_2$ and represents resin loss.

The weight percentage of glass fibre is:

$$\frac{M_2}{M_1} \text{ (in \%)}$$

and the weight percentage of resin is:

$$\frac{M_1 - M_2}{M_1} \text{ (in \%)}.$$

Epoxy resin prepregs

The same method is used. To determine the epoxy resin content, the volatile matter content of the resin must be taken into account. It is necessary to stove the specimen, initial weight M (NFT 57-570), in order to obtain the weight after all volatiles have evaporated, M_2. This material is then subjected to burn-off at 625 °C until only the glass fibre remains, mass M_3. The loss due to burn-off is $M = M_2 - M_3$ and represents resin loss. The weight percentage of resin in the prepeg is:

$$R = \frac{M_2 - M_3}{M_1} \text{ (in \%)}$$

and the weight percentage of volatile matter is:

$$T = \frac{M_1 - M_2}{M} \text{ (in \%)}.$$

The weight percentage of fibre is:

$$F = 100 - R - T \text{ (in \%)}.$$

Prepreg containing fillers
Standard NFT 57-518 (ISO 1172).

A burn-off method is used for preimpregnated mats or SMC, the protective films of which have been removed.

The organic fillers and resin are burnt off. The residue consists of glass fibre and mineral fillers. Hydrochloric acid is used to separate the mineral fillers from the fibre glass. After filtering only the glass fibre will remain. The test must be conducted on a minimum of two specimens.

The specimen is inserted into a thoroughly dry silica capsule weighing M_1 (g). The weight of the capsule and specimen together is M_2 (g). These are heated in an oven at 625 °C to constant weight, cooled in a desiccator and re-weighed, M_3.

This weight represents the total weight of the glass, the capsule and the non-combustible fillers SiO_2, CaO, $CaCO_3$,...). In practice non-combustible fillers account for the majority of filler material.

$M_3 - M_1 = \text{glass} + \text{non-combustible fillers.}$
$M_2 - M_1 = \text{initial weight of the sample.}$

The mass M_3 is reacted with concentrated hydrochloric acid to separate the glass fibre from the filler. The mixture is poured on to a sintered glass filter which must be dry and whose weight M_4 has been measured. The filter is rinsed and dried in an oven before being weighed, M_5. This represents the filter, the glass fibre and the fillers that did not dissolve in HCl (the latter being negligible).

$M_5 - M_4 = \text{glass} + \text{insoluble fillers.}$

The glass fibres are removed one by one with tweezers. What remains on the filter is the unburnt, insoluble filler, M_6.

The filler mass is $M_6 - M_4$. The filler ratio is

$$C = \frac{M_6 - M_4}{M_2 - M_1} \text{ (in \%)}.$$

The glass fibre weight percentage is:

$$F = \frac{(M_5 - M_4) - (M_6 - M_4)}{M_2 - M_1} \text{(in \%)}.$$

(b) Solution method

Standard NFT 57-608 (no ISO standard).

Prepregs containing no filler

Each sample with its protective film removed is placed in a container of mass M_1. The weight of the container plus sample is mass M_2. The container is filled with an appropriate solvent (acetone, methylene chloride, etc.) to dissolve the resin. The glass fibre is recovered on a sieve which has previously been weighed, mass M_3.

The sieve plus the glass fibre are dried in an oven and weighed, mass M_4. The mass of the specimen is $M_2 - M_1$; The mass of the glass fibre is $M_4 - M_3$. The glass fibre weight percentage is:

$$F = \frac{M_4 - M_3}{M_2 - M_1} \text{ (in \%).}$$

Prepregs and SMC containing fillers

An identical method is used. After dissolving the resin, the glass fibre and filler are recovered on a fine filter before being heated and dried:

$M_4 = $ filter + glass fibre + fillers;
$M_3 = $ filter mass.

The mass M_4 is reacted with concentrated HCl to separate the glass fibre from the filler. When the reaction is complete the glass fibre is washed and dried. The glass fibre has been removed from the filter, leaving only the insoluble filler. The filter is weighed, mass $M_6 = $ filter + filler; the mass of the filler is $M_6 - M_3$.

The weight percentage of filler is:

$$C = \frac{M_6 - M_3}{M_2 - M_1} \text{ (in \%).}$$

The weight percentage of glass fibre is:

$$F = \frac{M_4 - M_6}{M_2 - M_1} \text{ (in \%).}$$

2.3.5 Carbon fibre content

Solution method

Carbon prepregs do not generally contain any filler. Each specimen with its protective coating removed is weighed, mass M_1 (g). This mass is comprised of epoxy resin, R, and carbon fibres, F. So $M_1 = M_R + M_F$.

To dissolve the resin, the specimen is reacted with warm, undiluted nitric acid HNO_3 The carbon fibres are recovered on a weighed sintered glass filter, mass M_2 (g).

Thorough, careful washing will eliminate all traces of resin so that the filter will only retain the carbon fibre. The filter containing the fibres is then heated for 12 hours at 100 °C in order to dry the fibres. The filter and the fibres are then weighed very carefully, mass M_3 (g):

$$M_3 = M_F + M_2.$$

The fibre weight percentage is:

$$F = \frac{M_3 - M_2}{M_1} \text{ (in \%).}$$

Note that prepregs contain volatile matter. The determination of this is carried out as follows.

Take a specimen (with its protective film removed); weigh it, M_1 (g); heat for 12 hours at 100 °C; weigh again: M_2 (g).

The volatile weight percentage is

$$\frac{M_1 - M_2}{M_1} \text{ (in \%).}$$

A thermobalance can also be used.

2.3.6 Aramid fibre content

Specimens are snipped off, and protective films removed. Specimens are weighed: M_1 (g) $= M_R + M_F =$ resin + fibres.

To dissolve the resin the specimens are attacked with boiling tetrahydrofurane for five hours. The resin is then extracted with 'Soxhlet'.

The aramid fabric is recovered, air dried, stoved for 12 hours at 120 °C, and weighed, M_2 (g), when dry: M_F.

The fibre ratio is:

$$F = \frac{M_2}{M_1} \text{ (in \%).}$$

Note that the 'volatile' matter weight percentage can be obtained by heating the specimen for 15 minutes at 180 °C and weighing it. A thermobalance can also be used.

2.3.7 Plasticity of the prepreg

The resistance to flow of the resin through the fibres can be used to predict the behaviour of the prepreg when it is pressed. If the flow is too rapid, it can lead to areas of dry fibre (due to a lack of resin) in the artefact. If flow is too slow, the resin cannot permeate all the fibres so that porosity may be high. This can cause delamination.

Plasticity test
Standard NFT 57-513 (no ISO standard).

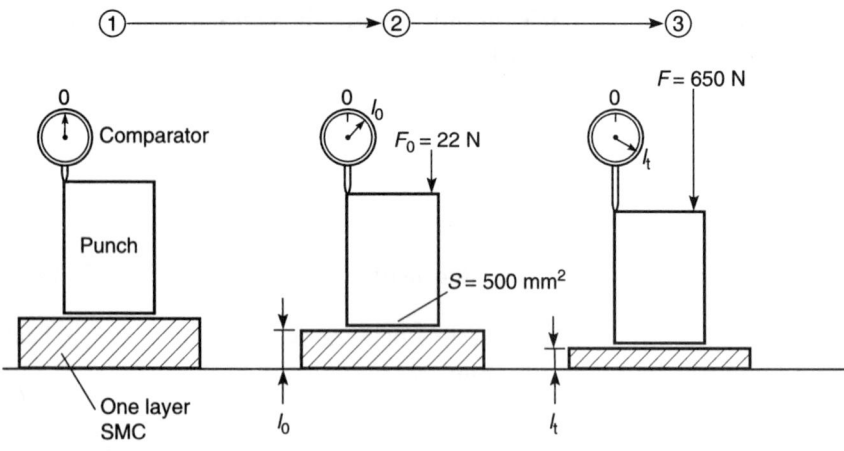

Figure 2.37 Plasticity test.

100 cm² specimens of prepreg or SMC mat are cut, their protective films retained, and they are subjected to a compressive load at room temperature using a flat punch whose area is 500 cm² as shown in Figure 2.37. The instantaneous thickness of a layer is measured with a gauge, and the force applied for a period of one to five minutes.

The test starts with an initial force F_0 of 22 N being applied for 30 seconds. The main force $F = 650$ N is then applied. The thickness of the specimen is measured at different times. At time t the plasticity is:

$$P_t = \frac{l_0 - l_t}{l_0} \text{ (in \%)}.$$

This is a measure of mouldability. The pattern of the plasticity curve as a function of time, known as the flow curve of the prepreg, can also be recorded.

Note that the above flow test can also be carried out on hot material if the conditions of hot moulding are to be reproduced. However, for the quality control of SMC upon delivery, room temperature testing is adequate.

2.3.8 Measuring the tack of a prepreg

This is a very important feature related to the processing of the prepreg. It is related to the adhesive qualities of the material and the ease with which two prepeg surfaces can adhere.

For processing it is essential that the prepeg has adequate tack. This quality must be reproducible and maintained throughout storage. It is often assessed

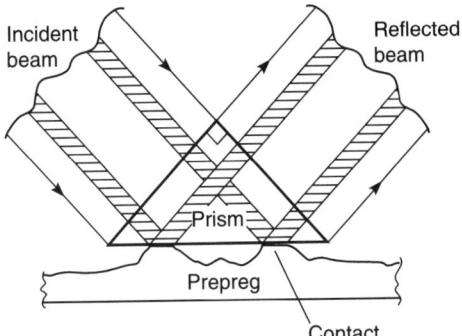

Figure 2.38 Measuring tack.

by touch, and there is no standard for this – only experience. Control devices called tackmeters do exist. A test piece is pressed against the prepreg, the rate of doing so, the pressure and duration of the test having been preselected. The tack is assessed by measuring the force required to remove the test piece.

An ingenious device, called the GIRA tackmeter, measures the contact area between the sample and the prepreg, using a prism allowing complete reflection of light, as shown in Figure 2.38. The prism is pressed against the prepreg and removed, the initial force applied having been selected in advance.

The surface of the prepreg is illuminated with a parallel beam of light at an angle of 45°. Where the prism and prepreg are in contact, the light is absorbed by the prepreg. Where the prism and prepreg are not in contact, the light is totally reflected.

The reflected light is measured with a photocell. This allows the measurement of the area, S, where the prism and prepreg are bonded at the moment when they are being pulled apart, and relates it to the maximum tensile force F_m. The resistance, σ, is the expression of the tack:

$$\sigma = \frac{F_m}{S} \, (\mathrm{N\,m^{-2}}).$$

2.3.9 Testing the reactivity of a prepreg

This is probably the most important test that is carried out on prepregs and on resins. If only one test is to be executed, it must be the reactivity test, for it shows:

- whether the resin has altered in any way; and
- whether the prepreg has aged before use (i.e. during storage).

In either case the use of the material must be reviewed or another set of curing parameters selected.

The reactivity test records the exotherm behaviour of a prepreg being cured at 120 °C. It is preferable to use a curing simulator (Vanhographe®, Kinemat® or again a DSC) for this test.

These devices provide the genuine thermal signature of the prepreg as if it were being cured in real life conditions.

(a) Standardized reactivity test on a prepreg

Standard NFT 57-516 (ISO 9780 relative to glass mats or SMC and prepregs).

The aim of the test is to record the exotherm curve of a prepreg being cured at a stable temperature $\theta = 120\,°C$ for polyesters and epoxies cured at 120 °C, or $\theta = 180\,°C$ for epoxies polymerized at 180 °C.

A thermocouple (iron–constantan) is placed between several layers of prepreg. The material is wrapped up in aluminium foil to give a perfectly airtight and waterproof package and the package is immersed in a polyglycol bath at $\theta\,°C$. The temperature increase inside the prepreg due to the polymerization reaction is recorded continuously by the thermocouple.

The length of time Δt needed for the prepreg to pass from $(\theta - 20)$ to $(\theta + 5)$ is calculated ($\theta\,°C$ is the temperature selected for the bath; the reactivity is expressed by Δt). This is illustrated in Figure 2.39.

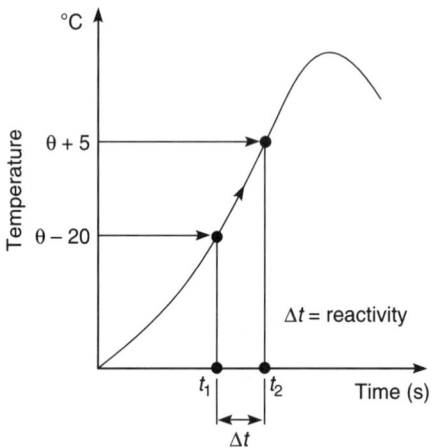

Figure 2.39 Reactivity curve.

(b) Test in curing simulator

The prepreg is tested by curing at the real processing temperature. Its stiffness is recorded as a function of time. Some devices bend the sample (Vanhographe®), others twist it (Kinemat®).

A thermocouple/recorder is placed in the prepreg and used to keep the oven at a constant temperature. During the curing process, the resin in the prepreg polymerizes and the structure becomes stiffer.

The Kinemat® applies a twisting force to the end of the sample twisting it by $\pm\alpha°$, as shown in Figure 2.40.

The sample E is a piece of prepreg measuring 50×15 mm, the ends of which are covered with aluminium foil to ensure that it remains between the clips and to avoid flow where the prepreg is stressed. At the other end of the sample, a force sensor indicates the instantaneous couple to which the sample is subjected and records it as a function of time during the isothermal curing process which is conducted under real curing conditions.

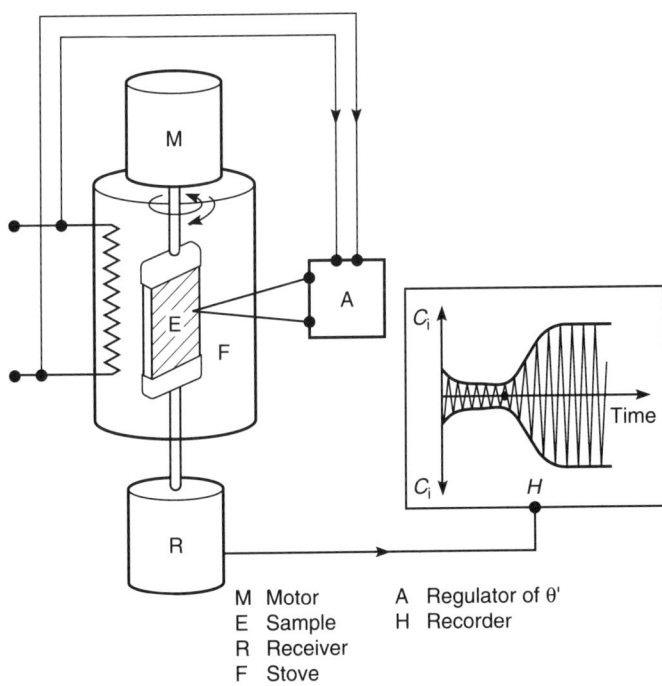

M Motor   A Regulator of θ'
E Sample   H Recorder
R Receiver
F Stove

Figure 2.40 Twister curing simulator principle.

The recording is symmetrical and constitutes a genuine signature of the prepreg, as shown in Figure 2.41. Four different zones can be distinguished:

- zone 1: the resin starts to flow under the action of heat;
- zone 2: solvents evaporate;
- zone 3: gel takes place and flow stops; when moulding, pressure must be applied shortly before t_1 = gel time = Gt;
- zone 4: the material is polymerized and does not undergo further changes.

It can now be demoulded.

It is thus possible to determine the gel time (Gt) and the polymerization time (t_{pol}) or demoulding time. The gel time, which corresponds to the beginning of the formation of spatial cross-linking in the resin, is one of the distinct features of the prepreg and can therefore be used to assess it.

It is strongly influenced by the type of resin used and the ageing of the resin.

It can be defined by the intersection of the tangent to the curve and the time axis. At this point the couple C_i is not zero, as shown in Figure 2.42.

Its value is $C_i(0)$, which is a distinctive feature of the prepreg. The difference $t_2 - t_1$ also characterizes the material.

All the above-mentioned values can therefore be used for quality control operations either on delivery of the prepreg or SMC or prior to using them after a long period of storage which might cause the materials to age.

Let $C_i(0)$ be the couple at gel time, Gt. Reactivity can be defined as the length of time necessary for the value of C_i to be x times that of $C_i(0)$; e.g. $x = 30$. This is shown in Figure 2.43.

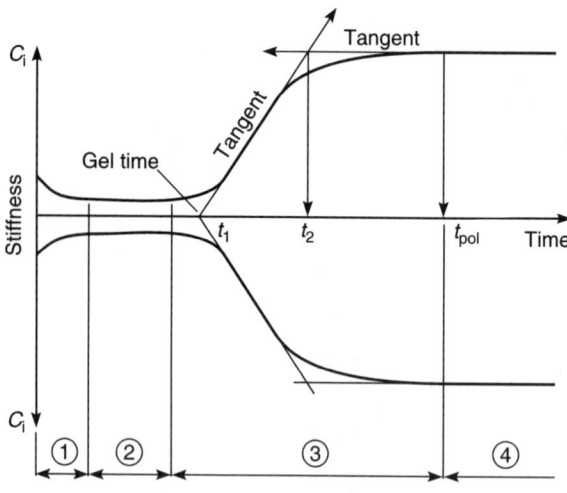

Figure 2.41 Curing simulation.

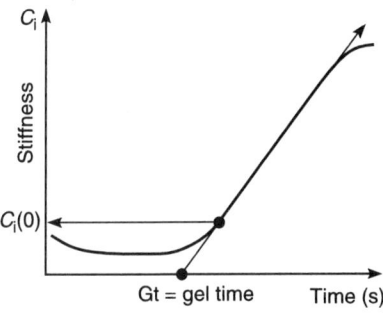

Figure 2.42

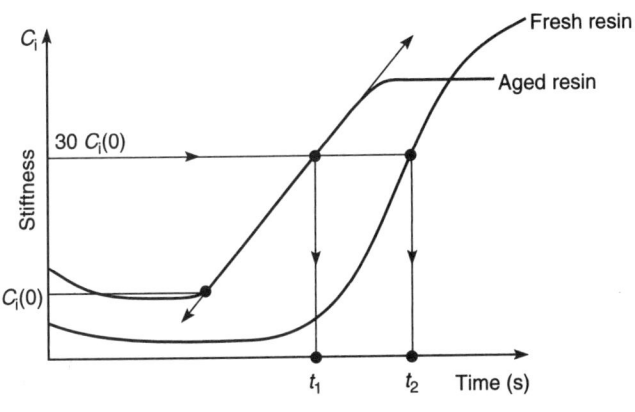

Figure 2.43

The reactivity of fresh resin is t_2; the reactivity of aged resin is t_1. An aged prepreg stiffens more rapidly when cured, because ageing has caused a considerable cross-linking to occur during storage.

(c) Testing reactivity using a DSC

The differential scanner calorimeter or DSC is undoubtedly the most useful device available to test the reactivity of a resin or to simulate curing. The DSC is a sensitive, easy-to-use device which only requires very small samples (a few milligrams). A detailed account of the test method is given in standard NFL 17-451.

An empty reference pan and a pan containing the sample of resin or prepreg are placed side by side in a small oven. The oven is programmed in such a way that the rise in temperature of the reference dish will be slow

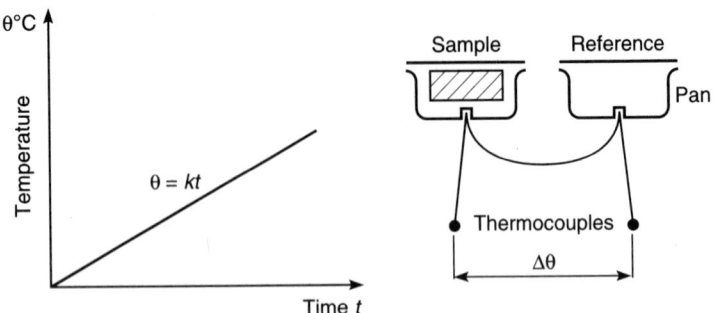

Figure 2.44 Heating cycle of the reference dish.

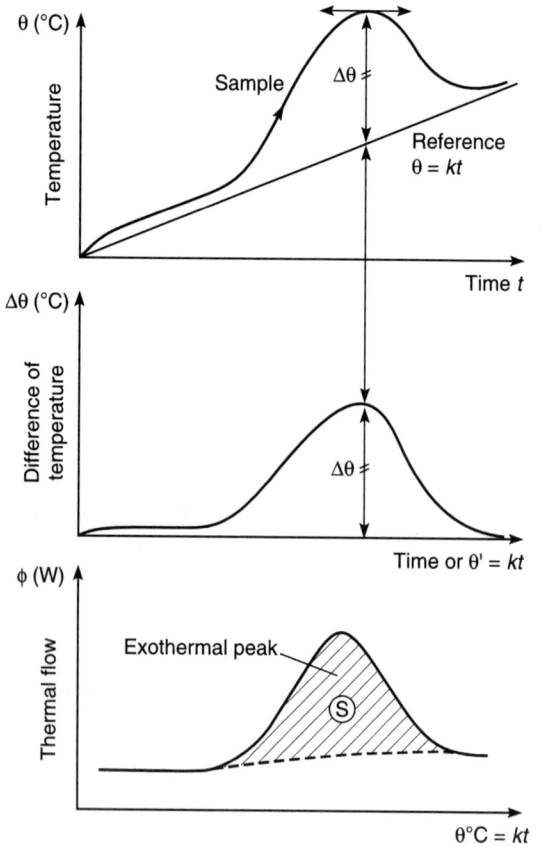

Figure 2.45 DSC principle.

and linear. The reference dish and the dish containing the sample are connected with a thermocouple allowing the temperatures of both dishes to be monitored at all times, as shown in Figure 2.44.

As the weight (m) of the sample and its specific temperature (c) have been measured accurately, it is easy to determine at any time the amount of heat flow ϕ lost or absorbed by the sample:

$$\phi = \frac{m \cdot c \cdot \Delta\theta}{\Delta t} \text{ (in watts).}$$

The device can record the evolution of the thermal flow ϕ with time and temperature, as shown in Figure 2.45.

It is thus possible to detect exothermal phenomena (e.g. polymerization) and endothermal reactions (e.g. evaporation of solvents).

To test the reactivity of a resin or a prepreg a small sample (a few grams) of prepreg is placed in a pan side by side with the reference pan, and heated slowly ($V = 2\,°C/min$). The heat flow generated by the slow polymerization process is recorded. The energy (w) lost is calculated (J/g, joules per gram, of material tested) by integrating the curve shown in Figure 2.46. The heat lost (W) is one of the characteristics of the material.

If the prepreg or resin has aged, i.e. if it has prepolymerized because of the time and temperature of storage, a DSC test will reveal a lower exotherm peak and energy loss (w), where $W' < W$, W being the energy lost on curing a fresh prepreg: see Figure 2.47.

The rate of ageing is W'/W (in %). It will also be noted that the temperature at which polymerization begins has changed from T to T'. Again ageing has caused it to rise.

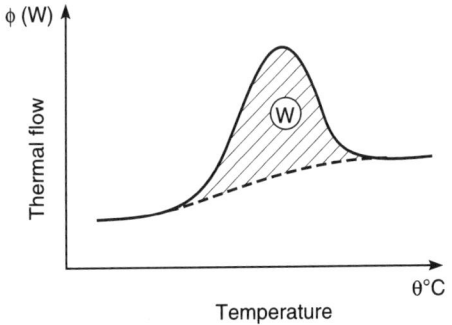

Figure 2.46

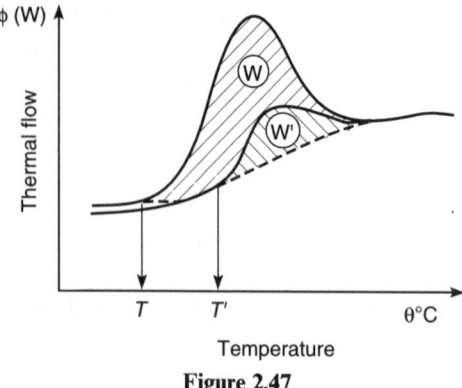

Figure 2.47

2.3.10 Rate of polymerization due to ageing

(a) Purpose

Prepregs polymerize slowly during storage. The rate of ageing can be assessed by measuring the fraction T of polymerized resin.

(b) Method: solution of non-polymerized resin

A sample of prepreg (mass M) is put into a suitable solvent in order to extract the soluble resin. After dissolution in a 'Soxhlet' and heating to eliminate volatile matter, what remains is mass M_1 (fibres + mineral fillings + residue of polymerized resin). Let M_F be the mass of fibres (non-soluble), M_C the mass of mineral filler and M_{RP} the mass of polymerized resin.
 Then:

$$M_1 = M_F + M_C + M_{RP}$$

and

$$M_2 = M_F + M_C$$

which, in the case of glass polyester prepregs, is determined by burning off the resin (NFT 57-557).
 In the case of carbon–epoxy prepregs the solution method can be used (see section 2.3.5).
 The same method when applied to a sample of fresh prepreg having the same mass M yields mass M' after solution and heating:

$$M' = M_F + M_C = M_2.$$

The soluble resin and volatile matter (which can be disregarded) have been eliminated.

The mass of non-polymerized resin M_0 contained in the sample of fresh prepreg is then:

$$M_0 = M - M'$$

(c) Rate of ageing

This can be expressed by the formula: $T = M_{RP}/M_0$ (in %). The suggested method is based on the method used to determine the amount of insoluble resin in glass polyester prepregs (old standard NFT 57-558).

2.4 SMC, BMC AND DMC CONTROLS

The industrial mass production of composite parts often requires the use of materials based on thermosetting resins, and mostly reinforced with E glass glass, known as BMC, DMC and SMC.

2.4.1 Definitions

Definitions are taken from standard NF ISO 8605 entitled: *Textile glass fibre reinforced plastics, sheet moulding compound (SMC) – basis for a specification.*

(a) Bulk moulding compound (BMC)

BMC is a moulding compound which is relatively homogeneous and composed of resin and a reinforcement of short fibres. This compound may contain inert fillers. BMC is used for compression moulding.

Cross-linking occurs under the combined action of both heat and pressure. BMC contains 1–2% in weight of thickening chemical agents called ripening agents.

(b) Dough moulding compound (DMC)

DMC is a relatively homogeneous moulding compound composed of resin and short fibres. It does not contain any ripening agent, but may on the other hand contain inert fillers to regulate viscosity. The glass fibres within DMC are often shorter than those in BMC in order to allow injection moulding without the risks of fibre breakage.

(c) Sheet moulding compound (SMC)

SMCs are moulding compounds in the form of continuous sheets between 1 and 25 mm thick. The sheets are not sticky. The compound is relatively homogeneous and is composed of resin, reinforcing fibres of various lengths, possibly inert fillers and a chemical thickening agent. Cross-linking can occur in SMC, under the action of heat and pressure.

These moulding materials are increasingly used in industry for the pressure moulding of various components including those for the automotive industry.

(d) Thick moulding compound (TMC)

TMC is a variety of SMC, with a thickness in excess of 25 mm.

(e) LS and LP materials (BMC, DMC, SMC)

The parts moulded from BMC, SMC and DMC are cheap to produce, and are used in many sectors including the car industry, the manufacturing of sanitary ware and household electrical appliances. These components must have accurate dimensions, and often, for example, for car bodies, have a very good finish. To improve the surface finish, thermoplastic antishrinkage agents are added (PE, PVAC, etc.).

LS or LP compounds can be defined as follows. LS (low shrink) materials are characterized by a low linear shrinkage which is generally no higher than 0.1%. LP (low profile) materials hardly shrink at all. Sometimes they even expand during moulding, giving a very smooth surface (class 'A' surface appearance for car bodies).

2.4.2 Composition of the compounds

No precise formulations will be given since each manufacturer varies the nature and percentage of the components depending on end requirement. The following results are averages obtained from analysing polyester-based products.

(a) BMC composition (by weight)

- polyester resin (UP + styrene)   15–25%
- short glass fibre (6–12 mm)   10–25%
- filler (calcium carbonate or kaolin or aluminium trihydrate)   40–60%
- LS or LP antishrinkage or expansion agents   5–15%
- polymerization initiator (TBTP)   0.4–0.6%
- ripening agent (MgO, CaO)   1–2%
- colouring pigments   1–5%
- demoulding agent (zinc stearate)   0.8–1.5%
- other additives (inhibitor, conduction agents, etc.)   0–5%

(b) DMC composition (by weight)

This is similar to the composition of a BMC, the only difference being that it does not contain any ripening agent.

(c) SMC composition (by weight)

• unsaturated polyester resin (UP + styrene)	34%
• fillers (CaCO$_3$, kaolin, etc.)	55%
• E glass fibre mat	30%
• LS, LP thermoplastic antishrinkage agents	5%
• TBTP initiator	1%
• ripening agent (MgO)	1–2%
• colouring pigment	2%
• demoulding agent (zinc stearate)	1%
• other additives:	the residue

2.4.3 Processing methods

Processing methods are as shown in Table 2.1.

Table 2.1

Material	*DMC*	*BMC*	*SMC*
Process	Injection	Compression	Compression
Pressure (in bars)	1000–1500	50–250	100–150
Mould temperature (°C)	150–200	120–170	≃ 150
Duration of cycle		According to	
(minutes)	≃ 1	thickness ⩾ 1	1–2

2.4.4 Controls on raw moulding compounds

(a) Introduction

Manufacturers deliver raw products in the following way. BMC and DMC are delivered in airtight barrels; SMC is delivered in sheets or rolls. It is protected by two removable sheets of polythene.

    These materials are subject to ageing during their storage life and the user is therefore advised to check them before use as they would with a conventional preimpregnated woven fabric. Because of the nature of automatic processing, any alteration in the composition or characteristics of the moulding compound could have very serious consequences resulting in inferior products or damaged machining.

(b) Possible quality checks

SMC, BMC and DMC are manufactured by reliable companies who provide the customer with the appropriate information regarding the quality and performance of their product. This information may be periodically checked by the user to see if the material has aged, etc. Among the controls are the following:

- resin ratio;
- filler ratio;
- glass fibre ratio;
- mass per unit area, or density;
- exotherm reactivity;
- viscosity;
- hot press mouldability;
- shrinkage;
- colour.

The mechanical properties of the material after moulding, i.e. flexural modulus and strength and impact strength can also be monitored to check out both raw materials and the fabrication process.

(c) Useful standards

Book 10 of the collection of AFNOR and ISO standards contains the main useful standards.

Standards for SMC
These often cover other glass reinforced polyester materials.

- NFT 57-601 and ISO 10352. Determination of mass per unit area: prepregs and SMC.
- NFT 57-570. Preimpregnated mats. Conventional volatile matter content (styrene).
- ISO 9782 or NFT 57-602. Prepregs: determination of conventional volatile matter (styrene).
- NFT 57-513 or ISO 3616. Determination of plasticity of preimpregnated mats. Mats: thickness under load and recovery.
- NFT 57-514 or ISO 4900. Mouldability in hot press – prepregs.

- NFT 57-515. Measurement of expansion.
- NFT 57-516 or ISO 9780. Conventional reactivity of resin pastes and prepregs.
- NFT 57-608. Prepregs: fibre and resin content by dissolving the matrix.
- NFT 57-557. Prepregs glass fibre fabric: determination of loss through burn-off.
- NFT 57-571. Preimpregnated mats. Determination of ignition loss.
- NFT 57-603. Prepregs; conventional resin flow rate.
- NFT 57-518 or ISO 1172. Glass and filler content by burn-off.
- NFT 57-519. Preimpregnated mats. Determination of shrinkage during compression moulding, and post shrinkage.

Standards for BMC and DMC
- ISO 8605. BMC and SMC. Basis for specification.
- ISO 293 and NFT 58-003. BMC, DMC compression moulding test.
- NFT 51-401 (ISO 2577). Determination of shrinkage and post shrinkage for thermosetting plastics.
- NFT 58-008. Injection moulding (DMC) determination of characteristics.
- NFT 07-009 (ISO 6989). Fibre length: distribution.
- NFT 51-402. Determination of transfer flow.

2.4.5 Controls for SMC

(a) The problem

The properties of SMC show considerable scatter. The main reason for this is that it is impossible to achieve a good dispersion of small quantities (i.e. few percentage points) of additives such as ripening agents, polymerization initiators, etc. within the compound. Dispersion depends on the type of mixing machine used and on the process employed to impregnate the glass mat reinforcement.

A 1 m wide roll of SMC will show noticeable scatter in additive control both across the width and down the length.

(b) Mass per unit area (see Section 2.3.2)

(c) Glass fibre content (see Section 2.3.4)

(d) Reactivity on exothermal flow (see Section 2.3.9(a))

(e) Plasticity (see Section 2.3.7)

(f) Filler content (see Section 2.3.4(b))

In addition to the checks carried out on prepregs, the following should be monitored: compression behaviour, moulding shrinkage and mouldability.

(g) Compression moulding shrinkage

This is determined by measuring, as accurately as possible (after moulding and cooling at room temperature), the difference in diameter or dimension between the moulded part D_1 and inside the mould, D_0.

Both the mould and the punch are made of metal. A minimum of three 3–5 mm thick specimens are moulded. The pressure, P, the temperature θ and curing time must be the same as those used in factory production, as shown in Figure 2.48.

The linear shrinkage due to compression moulding, R_m, is given by.

$$R_m = \frac{D_0 - D_1}{D_0} \text{ (in \%)}.$$

Prior knowledge of R_m is very useful for mould design. It allows the user to decide whether low shrink (LS) or low profile (LP) material is required. For further details, see NFT 57-519 and NFT 57-515.

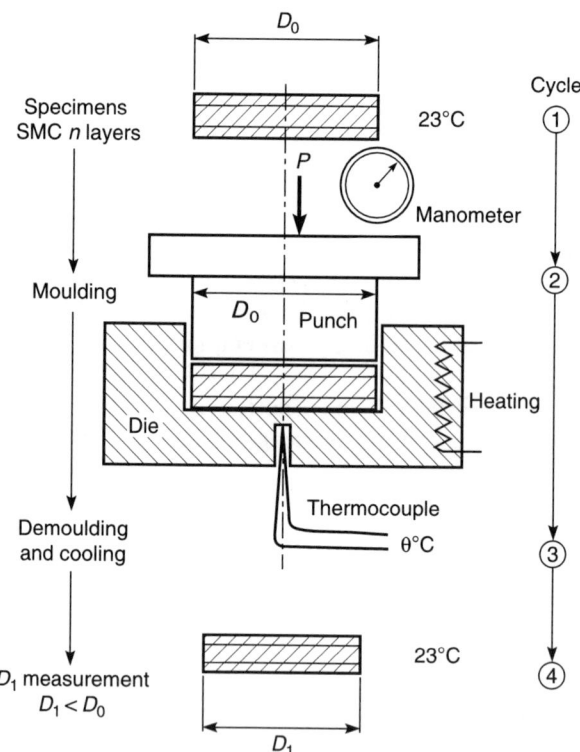

Figure 2.48 Measurement of shrinkage.

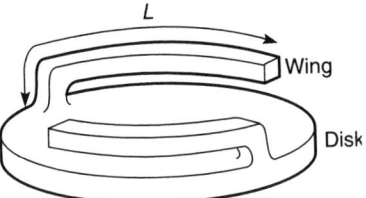

Figure 2.49 Specimen of mouldability.

(h) Mouldability

In the course of compression hot-moulding, the SMC must flow to fill the entire inside of the mould without fracturing or leaving areas deficient in material. This characteristic, which is called mouldability, can be evaluated by moulding a specially shaped test specimen under the same conditions of pressure, temperature and time as are used in the production. For example, in a test based on the AFNOR standard NFT 57-514 and ISO 4900 a disc of uncured SMC is forced by a punch into a heated cylindrical mould (similar to that used for shrinkage tests, but with grooves on the side). The punch forces the disc-shaped charge of SMC to flow into the two grooves in the side of the mould, as shown in Figure 2.49. The specimen obtained thus has two protruberances whose length *L* increases with the mouldability of the compound.

2.4.6 Essential tests on BMC or DMC

There are few if any standards for these materials. However the following tests are useful:

- determination of the density of the compound before compression;
- mouldability in a hot-moulding press (NFT 57-514 and ISO 4900);
- reactivity (NFT 57-516 and ISO 9780);
- glass and filler content by a burn-off method (NFT57-518 and ISO 1172);
- glass and resin content by dissolution method NFT 57-608;
- compression moulding test NFT 58-003 or ISO 293;
- injection moulding. Determination of characteristics by NFT 58-008.

The mechanical properties of the pieces moulded by compression of BMC or by injection of DMC will depend on:

- the length of the glass fibres;
- their dispersion in the compound (this may be more or less homogeneous);
- their surface treatment (size chemistry).

(There are no standards for such tests!)

Testing the linear shrinkage of thermosetting compounds during compression casting consists of compression moulding (using the same P, θ, t cycle as is employed in mass production) of parallel sided bars of BMC in a metal mould. The length of the inside of the mould is L_0. The length of the moulded but cooled bar is L_1. As $L_0 > L_1$, the linear shrinkage R_m is:

$$R_m = \frac{L_0 - L_1}{L_0} \text{ (in \%)}.$$

Standard: NFT 57-519.

2.4.7 Compression moulding testing of SMC and BMC

(a) Nature of the problem – quantities to be measured

The compression moulding of SMC and BMC is the most widespread form of fabrication used for glass fibre polyester or glass fibre epoxy composites.

Since it is now used to manufacture car components, excellent quality is required both from a mechanical point of view (technical parts) and from the point of view of surface appearance (surface appearance Class-A for car body parts). It is therefore necessary to perform quality control tests in real life conditions of moulding in a factory. This necessitates the use of a moulding press fitted with the appropriate sensors.

The moulding procedure for an SMC piece is as follows:

- introducing the SMC (or BMC) into the hot mould (temperature about 150 °C);
- closing the mould ($V \simeq 2$ mm/s) and applying pressure (about 100 bars);
- curing;
- opening the mould;
- ejecting the piece.

The following occurs in the mould:

- the SMC heats up through contact with the mould;
- pressure is applied;
- the charge flows into every part of the mould;
- the air contained in the mould is forced out;
- curing begins in the areas which have reached the required temperature;
- curing takes place throughout the compound (heat is released and shrinkage occurs);
- compensatory swelling occurs because of LS and LP anti-shrinkage agents.

The parameters to be checked are the following:

- heat flux within the mould;
- orientation of the reinforcement in the compound;

- exothermal reactivity;
- shrinkage during cure;
- shrinkage compensation (both mechanical quality and surface appearance will depend on it).

(b) Test using a mould fitted with sensors

It is difficult to check that the mould has been filled properly using sensors. It is easier to examine the processed component after fabrication. The parameters to be checked with the mould in the press are:

- reactivity or exothermal reaction;
- shrinkage;
- antishrinkage effect.

For that purpose, the press and the mould will be equipped with the following, shown in Figure 2.50:

- a speed sensor on the counter mould;
- temperature sensors on the mould or in the moulded part;
- press sensor on the press;
- a sensor on the counter mould to monitor both the position of the compound and its movements.

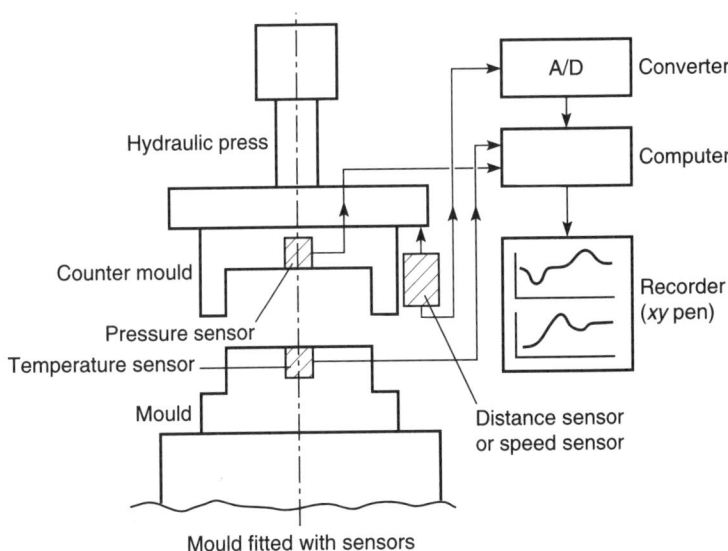

Figure 2.50 Test of mouldability.

Readings of thermal or volume changes taking place within the compound during the moulding process can thus be obtained.

Suppose for instance that the clamping force F_m of the mould were under servo-control and were maintained at a constant level $F_m \simeq$ const through acting solely on the lock of the mould. The x and y readings used for analysis would be the following:

- part (or mould) temperature T in time t;
- position D of the counter-mould in relation to the mould;
- pressure bearing on the mould or clamping pressure F as a function of time.

We can see from Figure 2.51 that on curve (T, t) (a) the temperature of the mould drops when in contact with the cold charge (1). The temperature of the mould rises until it again reaches θ (2). The exotherm reactivity due to curing is shown at (3). There is a progressive return of the mould to temperature θ (4).

On curve (F, T) (b) we see the thermal expansion caused by the heating of the compound in contact with the mould (1). As a result F rises to a

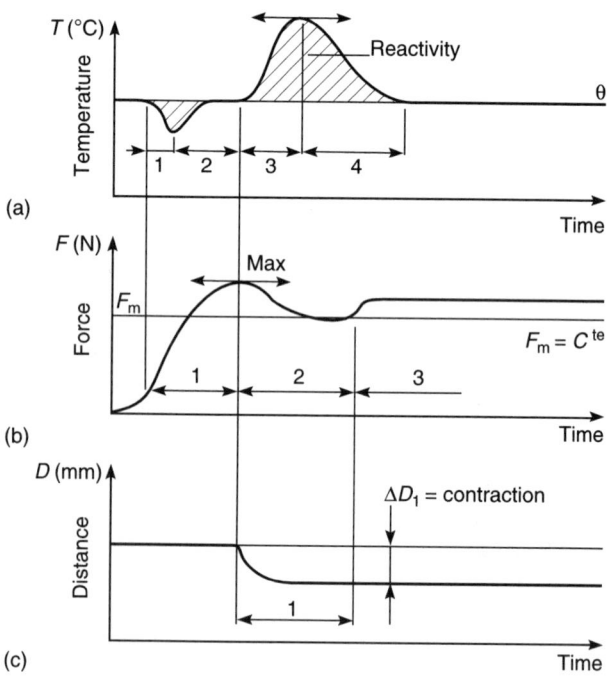

Figure 2.51 Constant clamping pressure moulding.

maximum F_{max} because the mould remains closed. Shrinkage in the course of curing causes a drop in pressure, but as the mould automatically closes further, pressure is maintained at level F_{max} (2). The effect of swelling agents that may have been added must be taken into account. As there is no pressure release mechanism, the press control does not correct the pressure (3).

On curve (D, t) (c) there is shrinkage ΔD_1 due to curing. The mould closes further automatically to raise the pressure back to level F_m. Swelling is detected with a movement sensor.

The main drawbacks of this method are the following:

• It is necessary that a press similar to those used in the production line should be used in the laboratory.
• It is necessary to fit the equipment used with sensors every time a new mould is used.
• It is difficult, when large pieces of equipment are used, to measure slight shrinkage or swelling.

For a 2 mm piece, shrinkage can be estimated at 20 to 10 µm, and swelling at 0 to 20 µm.

This approach, although it makes it easier to fine tune the production of real-size components, remains too costly for quality control checks on SMC or BMC compounds. For the latter, small test pieces and laboratory equipment should be used (moulding simulators).

(c) Check with moulding simulator

Since 1970, researchers have been studying the shrinkage of polyester compounds (Bartkus, Mitani, Kinkelaar, Kau). The equipment used in their experiments comprises a small mould filled with BMC, with pressure applied through a piston. The parameters which are measured are the temperature and expansion or contraction of the compound. The mould is calibrated in an isothermal chamber.

The Plastoreactomat® (Insa-Prodemat France) appeared in 1990 and is illustrated in Figure 2.52. It is a BMC moulding simulator whose advantages are the following:

• It recreates conditions close to those found in the production line. However, it only requires the use of small test pieces ($\phi = 20$ mm, $e = 1$–4 mm).
• It provides information about the intrinsic properties of the material – shrinkage compensation, exothermal reaction.
• During the production process of a BMC, DMC, SMC or prepreg, it allows the operator to check the moulding compounds before the addition of fibres as well as during the ripening process.
• It makes it possible to reject unsuitable batches of BMC and SMC.

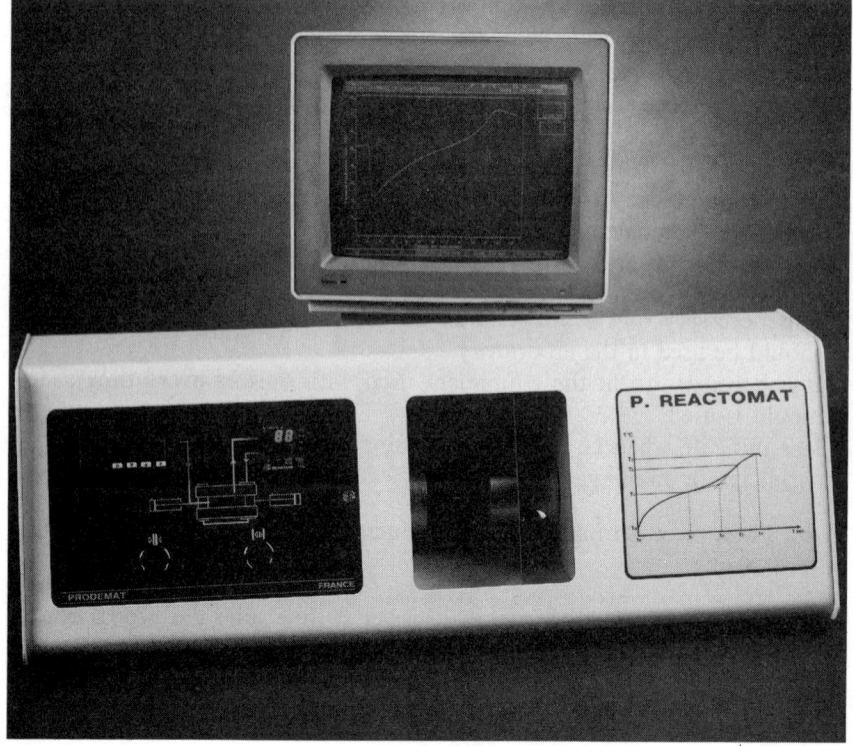

Figure 2.52 Plastoreactomat®.

The device has a piston linked to a mobile strut. The piston is driven into a small cylindrical mould ($\phi = 20$ mm) containing the SMC or BMC or DMC. The temperature of the mould is raised (from 20 to 200 °C). The pressure exerted by the piston which determines the moulding force F_m on the material is selected to fit industrial requirements.

Three measurements are taken in the course of moulding as shown in Figure 2.53:

- the force F exerted by the piston on the material;
- the temperature of the material;
- the position or movement A of the piston in relation to the bottom of the mould.

The device can be used as follows: either the moulding force F_m is fixed at a set load or the displacement A is set at a fixed level – though in this case the piston ceases to move if F_m is reached before the set displacement.

Suppose the device were used with fixed displacement. The readings used

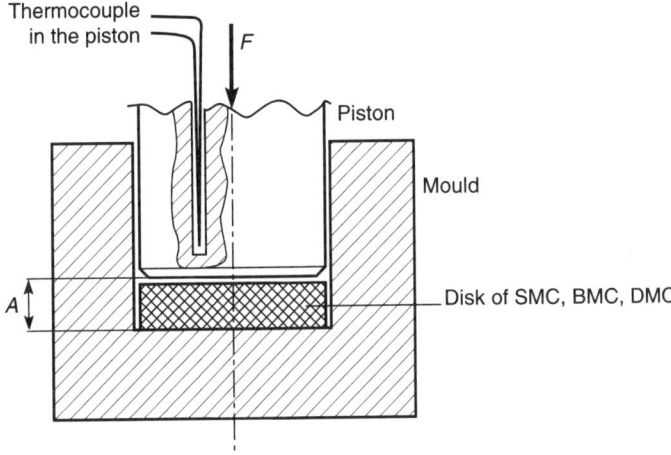

Figure 2.53

to analyse the phenomena occurring during the moulding process are temperature T as a function of time and force F as a function of time. The thickness of the test pieces remains the same during the test.

We can see from Figure 2.54(a) that on curve (T, t) there is a drop in temperature when cold material is introduced into the mould at θ (1). An exothermal peak due to curing is followed by a return to the moulding temperature θ (2). On curve (F, t) (Figure 2.54(b)) there is a rapid rise to nominal force F_m at which point the mobile strut and the piston stop moving (1). Thermal expansion of the material in contact with the hot mould causes a rise to maximum force F_{max} (2). Curing shrinkage corresponds to the drop ΔF_1 in force (3). Compensation for the shrinkage by antishrinkage agents (LS or LP) causes a rise ΔF_2 in force because the piston cannot move back to ease the pressure (4).

The Plastoreactomat® can also be used differently, with the moulding clamping force having been set in advance and regulated throughout moulding. In this case, the readings are the same as when moulding is performed at constant pressure (see the above description of a mould fitted with sensors).

Some results obtained from tests on SMC

Influence of moulding temperature on shrinkage In industry, moulding is usually carried out under isothermal conditions. The influence of moulding temperature θ on shrinkage and shrinkage compensation can be studied by selecting the thickness of the piece to be moulded, e mm, and maintaining

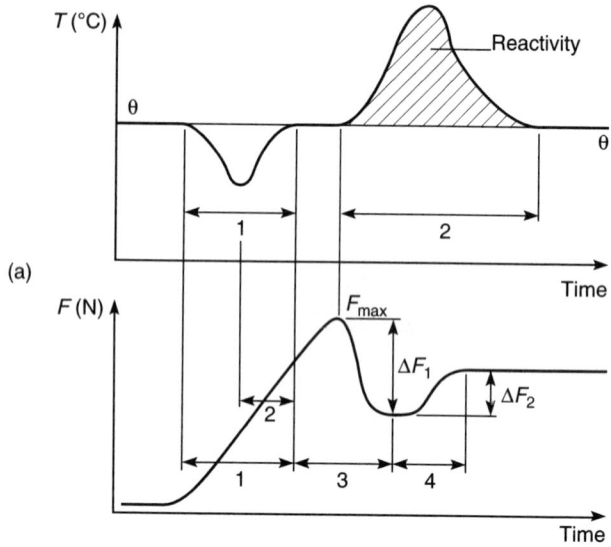

(a)

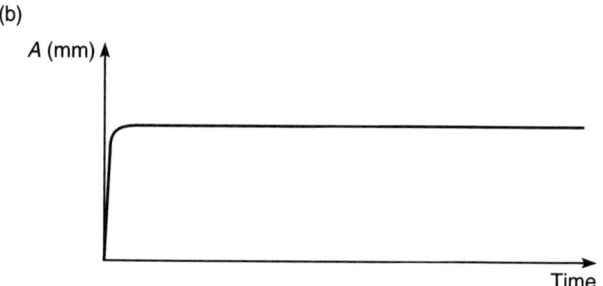

(b)

Figure 2.54 Moulding when mould remains clamped in the same way throughout moulding.

clamping pressure at a constant level. Shrinkage is expressed by $\Delta F_1/ek$; swelling by $\Delta F_2/ek$, k being the mechanical stiffness of the device. The result given in Figure 2.55 was obtained by L. Jayle (Insa-Lyon).

The overall shrinkage is equal to $(\Delta F_1 - \Delta F_2)/ek$. A rise in moulding temperature thus entails a decrease in overall shrinkage.

Influence of moulding pressure on shrinkage Other tests carried out on SMC have shown that an increase in moulding pressure (force) did not lead to swelling at the end of the curing process. The risk that the surface appearance of the component might be of poor quality was therefore high.

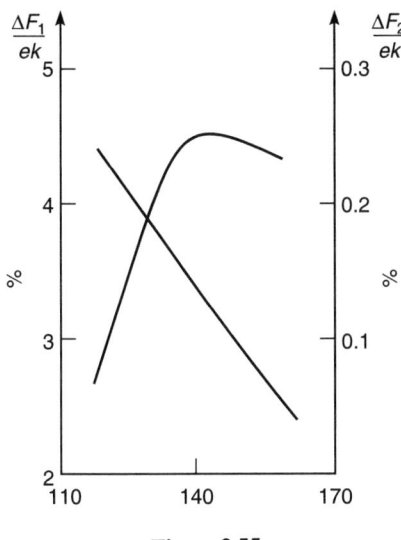

Figure 2.55

Influence of storage ageing on shrinkage The ageing of an SMC can be followed by studying its plasticity $P(t)$ during its storage life at room temperature. The check on plasticity is performed according to standard NFT 57-513 (see Section 2.3.7). The curve we get for $P(t)$ as a function of time is shown in Figure 2.56 and has three distinctive zones:

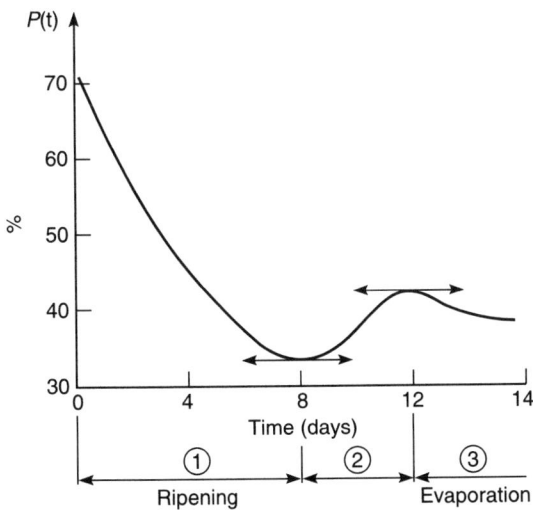

Figure 2.56

- Zone 1. From the first to the eighth day, plasticity decreases owing to the ripening of the product.
- Zone 2. From the eighth to the twelfth day, plasticity increases. This might be a consequence of ambient moisture.
- Zone 3. Beyond the twelfth day, plasticity starts decreasing again because of the evaporation of the styrene contained in the polyester resin.

2.4.8 Control of DMC through injection moulding

Some manufacturers, for instance Orkem (France), have developed an injection press fitted with sensors, the aim of which is to predict and check the behaviour of DMCs when moulded.

The following are checked:

- mouldability (rheology);
- shrinkage;
- reactivity.

(a) Interaction between parameters

This check is carried out under industrial conditions and aims at guaranteeing minimal cycle lengths and shrinkage consistency by allowing adjustment of the moulding parameters if the DMC has, for instance, aged. The cycle length depends on the speed at which the mould is filled as well as on the speed of cross-linking.

The speed at which the mould can be filled depends on the viscosity of the cold compound when it comes out of the injection nozzle, and also on its viscosity once it warms up in the mould. Viscosity also depends on injection pressure, moulding temperature and composition of the compound (glass content, length of fibre, fillers, etc.). Cross-linking depends directly on the composition of the resin and moulding temperature. Shrinkage is dependent on the LS and LP additions added to the blend.

The interactions that occur between the different parameters are too complex to be studied separately. It is better to characterize in the mould itself the kinetics of curing and the shrinkage, and then to use the test pieces for mechanical testing (flexural properties, impact etc.).

(b) Equipment needed

The press must be fitted with a certain number of sensors and controls so that reproducible measurements can be made.

- The regulation of temperature must be possible.
- The regulation of injection flow must be possible.

- Pressure sensors must be fitted on the injection nozzle and in the mould.
- Temperature sensors must be fitted in the mould.

All the data collected is processed by a microcomputer.

(c) Mould and test pieces

There are currently no standards defining what the mould should be like. According to Orkem, the mould should consist of three parts, the shapes of which would be representative of three types of pieces (linear, flat and massive) as shown in Figure 2.57:

- a spiral whose section would be rectangular;
- a small plate (100 × 50 × 4 mm);
- a cylindrical vat (10 cm³).

(d) Principle measurements made

Viscosity is measured on the cold compound. This measurement is made at the injector nozzle, with the mould open, as shown in Figure 2.58. Two pressure sensors P_1 and P_2, ΔZ apart, and a temperature sensor are fitted on to the nozzle of radius r. If the nozzle has a constant flow Q we get:

shear gradient:
$$\gamma = \frac{3Q}{\pi/2}$$

shearing strain:
$$\tau = \frac{P_1 - P_2}{\Delta Z} \times \frac{r}{2}$$

and apparent dynamic viscosity $\eta = \tau/\gamma$.

The vat-shaped part of the mould is fitted with a thermocouple which

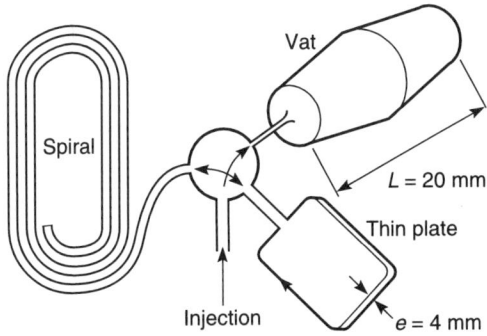

Figure 2.57 DMC test piece injection moulding (after Cray-Valley)

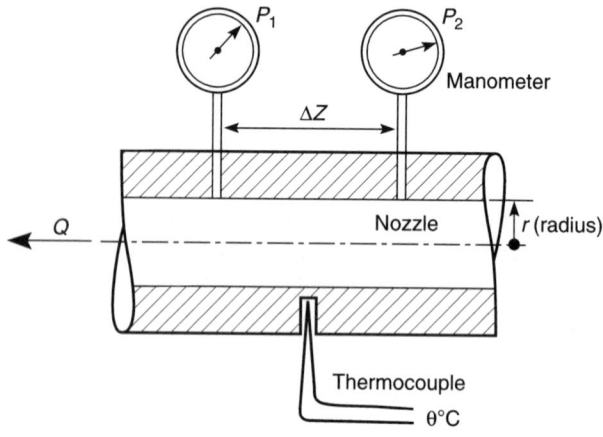

Figure 2.58

gives the temperature in the centre of the compound as a function of time during moulding. The temperature of the mould remains more or less constant, as shown in Figure 2.59.

Shrinkage can be measured on the small plate by comparing marks on the cold test piece with those on the cold mould, as in Figure 2.60.

$$R_L = \frac{L_M - L_P}{L_M} = \text{longitudinal shrinkage ratio}$$

$$R_T = \frac{T_M - T_P}{T_M} = \text{transverse shrinkage ratio}$$

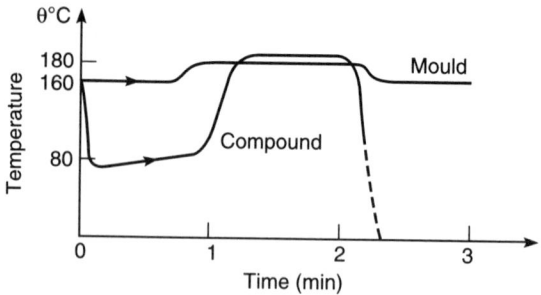

Figure 2.59 Reactivity measured on vat-shaped mould.

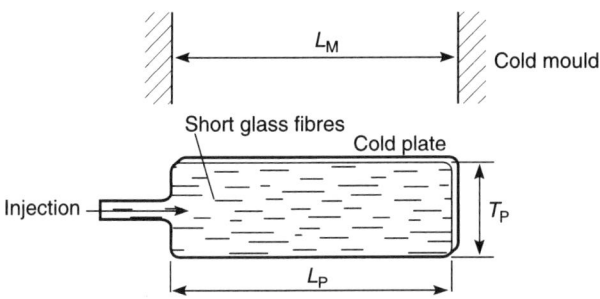

Figure 2.60 Longitudinal shrinkage ratio.

where L_M, L_P, T_M, T_P refer to the cold mould and cold plate. Shrinkage is greater lengthwise (in the direction of injection flow) than crosswise. This phenomenon is due to the fact that short reinforcement glass fibres orient themselves naturally lengthwise when they are injected. Crosswise, the fibres limit shrinkage or swelling.

The antishrinkage effect that occurs in low-profile DMC can be assessed by measuring the pressure P in relation to moulding time t in the vat-shaped test piece, as shown in Figure 2.61.

On the curve we can see:

1. A rapid rise in pressure during injection.
2. The injection nozzle clogs up because the compound cross-links in the

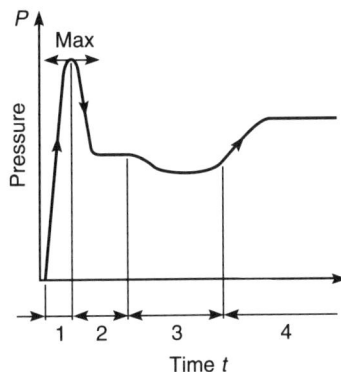

Figure 2.61 Pressure in the vat-shaped mould during injection.

thin canal. The DMC contained in the vat is thus isolated from the press and pressure therefore decreases.
3. A slight drop in pressure due to the shrinkage of polyester resin.
4. A rise in pressure caused by thermoplastic swelling agents.

(e) Cases in which a press and mould fitted with sensors can be used

The mould described above makes possible

- the elaboration of new DMC and
- checks upon delivery

without it being necessary to carry out moulding tests in real life conditions – making a dozen test pieces by injection moulding does not take much time (statistical measurement). The drawback is that such a mould is a costly piece of equipment which diverts a moulding press from industrial use. That is the reason why it is mainly used by the manufacturers of moulding compounds rather than by their users.

By using this kind of technology, it becomes possible to study:

- the formulation of DMC (influence of fillers and antishrinkage agents);
- the evolution of the product during storage life (ageing);
- the moulding cycle (pressures, temperature, time).

(f) Sample results (Orkem Society)

Comparing the reactivities of a so-called fast DMC and a slow one, this method gives more accurate results than the standard test (Section 2.3.9(a)) as in Figure 2.62: the shorter the curing cycle the greater the productivity.

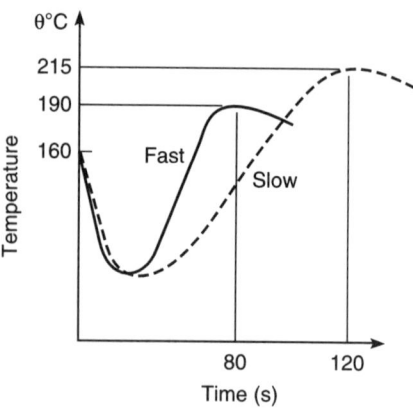

Figure 2.62 Comparing reactivities.

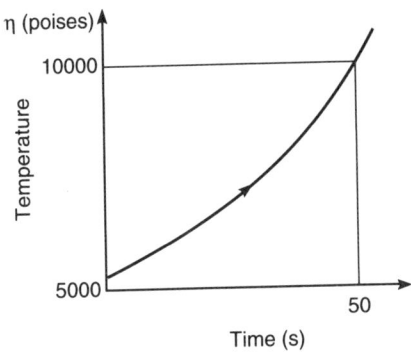

Figure 2.63

A sizeable increase in dynamic viscosity η occurs due to ageing. This can be detected by measuring, at room temperature, the viscosity of the DMC charge as it goes through the injector nozzle, as in Figure 2.63.

Studying shrinkage or swelling can be done on plates injection moulded from LS or LP material. An increase of pressure in the mould leads to swelling giving an excellent surface appearance.

2.5 CONTROL OF SANDWICH CORES

2.5.1 Sandwich structures definition – pros and cons

Sandwich core materials enable very rigid beams with thin skins to be manufactured. Individual components can be joined adhesively to produce frames, structures and lattice work of high rigidity and low weight. A sandwich beam consists basically of two monolithic laminated skins adhesively bonded to either side of a light core. The beam has a large moment of inertia about the natural axes, xx'.

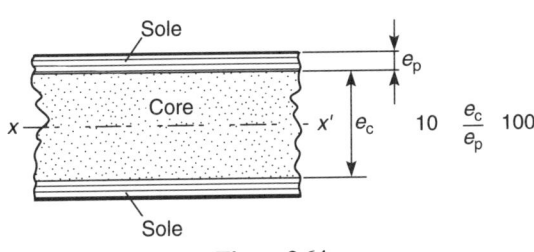

Figure 2.64

The laminated skins have good mechanical characteristics compared with the core which is mechanically weak. In addition the core gives good thermal insulation properties. Figure 2.64 shows a typical sandwich structure.

The main drawbacks are susceptibility to buckling and lack of resistance to penetration from sharp objects.

2.5.2 Different types of sandwich cores

The materials used to make the cores include the following:

- aluminium alloys: series 5000, 2000, 3003 ASTM;
- balsa wood;
- resin impregnated paper;
- E glass woven fabric impregnated with (UP) polyester resin;
- Nomex® (Du Pont de Nemours) aramid woven fabric impregnated with phenolic resin (PF) – sometimes in the form of a honeycomb;
- polyethane (PE);
- expanded polystyrene (PS); extruded PS;
- polypropylene (PP);
- expanded polyurethane (PUR);
- expanded polyvinylchloride (PVC);
- Firet Coremat® (Netherlands) (Coremat® is a mat of non-woven polyester fibres and 50% of thermoplastic (PVDC) microballs linked with a polyacrylic material. It is 1 to 5 mm thick and can be impregnated with resin);
- polyacrylic materials.

One can distinguish various types depending on how they are made:

- Solid cores (actually porous): plywood, balsa wood, foams (acrylic, PS, PUR, PVC, PE), blend of microballs and resin binder. Density varies.

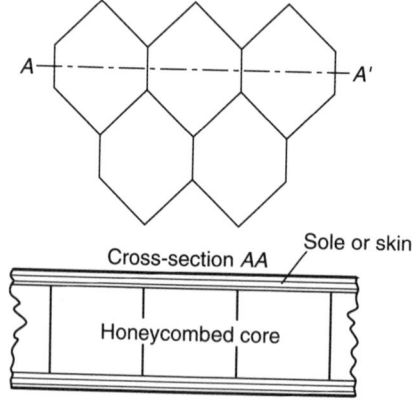

Figure 2.65

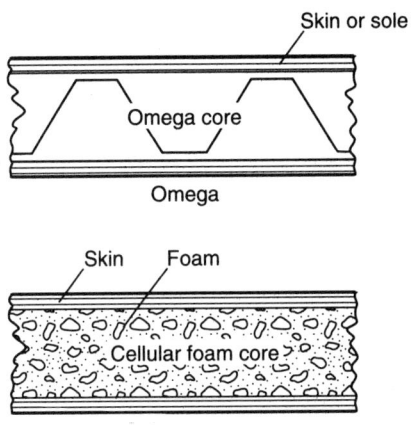

Figure 2.66

- Honeycombed cores, illustrated in Figure 2.65, made of aluminium alloys, impregnated paper, impregnated Nomex® paper, impregnated glass fibre woven fabric, polypropylene, aramid (used in aerospace industry).

Note that aluminium and Nomex® honeycombs have superior mechanical characteristics to those of foams, shown in Figure 2.66, and low densities (30–$80\,\text{kg/m}^3$). It is for this reason that they are so widely used in the aviation and aerospace industries.

Structures containing microballs (glass or phenolic) and resin are widely used owing to their high resistance to crushing and range of densities, depending on the ratio of microballs to resin.

2.5.3 Stresses and strains on the core

The main loading modes are buckling; tension, compression; shearing; and punching.

Tension–compression
When a sandwich beam is subjected to flexural strain, as shown in Figure 2.67, tension and compression strains are equivalent and mainly restricted to the thin laminated skins. However, the core must be able to resist tensile and compression forces generated in the adhesive interface.

Local loading
The core may be deformed locally in compression by the application of point-loads on the skins, as illustrated in Figure 2.68. When honeycomb composite

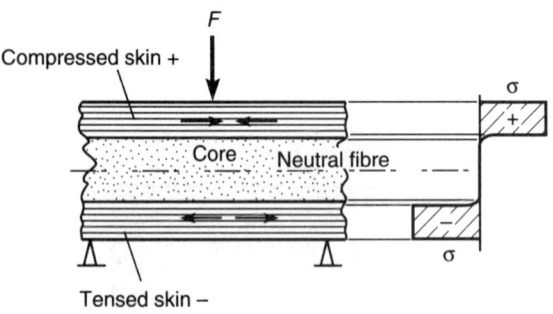

Figure 2.67 Sandwich subjected to flexural strain.

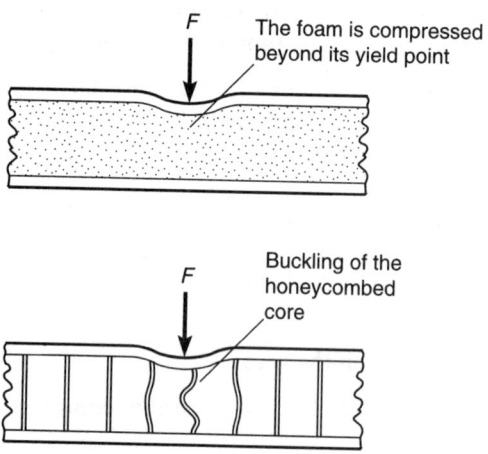

Figure 2.68 Ramming effect.

sandwiches were first used to make aeroplane floors, there were problems with puncturing by stiletto heels. Thus the core must be compression resistant.

Buckling
The outer skins of the beam are subject to buckling owing to the low compression resistance of the core.

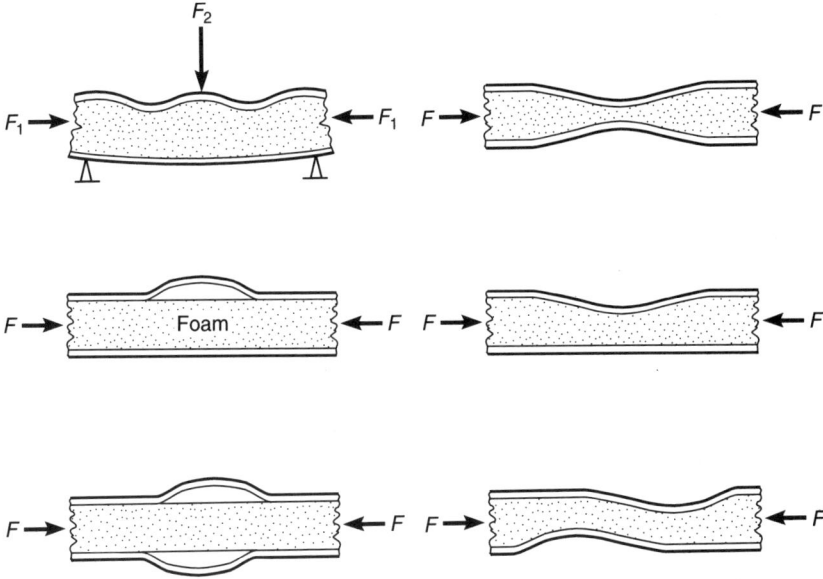

Figure 2.69 Different types of buckling (after D. Gay)

Examples are shown in Figure 2.69. The skin–core interface may even separate.

Shearing
In some loading situations, the core is subjected to a shear stress strain τ as is the adhesively bonded interface. The core requires a certain shear strength to avoid failure.

(a) Shear stress estimation

If a beam, supported at two points, is loaded centrally by a force F, as shown in Figure 2.70, the resulting reactions from the supports will be the following:

$$R_A = R_B = F/2.$$

Shear stress $= T = F/2.$

For a sandwich beam of overall thickness h with thin skins of thickness e, the shear stress τ can be assumed to be taken evenly by the core. Thus it follows that

$$\tau \simeq \frac{T}{b(h - 2e)} \quad \text{if } e \ll h$$

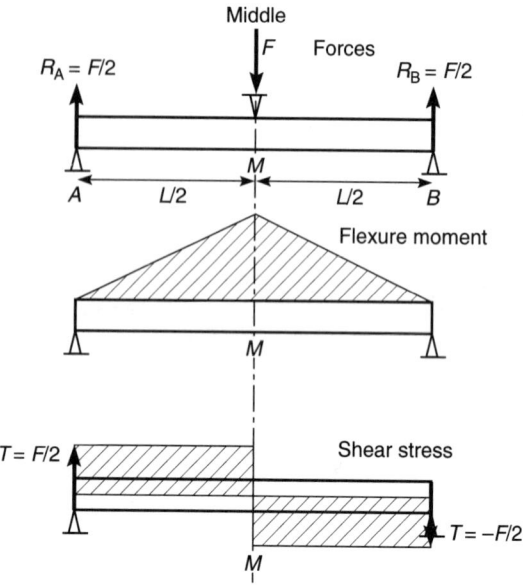

Figure 2.70 Three-point flexure test.

but $T = F/2$ hence

$$\tau \simeq \frac{F}{2b(h - 2e)}.$$

In an alternative approximate method for an isotropic beam, the maximum value of shear stress τ which occurs at the neutral axis is:

$$\tau_{max} = \frac{3}{2} \frac{T}{b \cdot y}$$

where y is the height of the beam and b is the width of the beam.

Apply this relation to a sandwich core which is subjected to a shear load T. Let $y = h - 2e = c$. Then

$$\tau_{max} = \frac{3}{2} \times \frac{T}{b(h - 2e)};$$

but $T = F/2$, hence:

$$\tau_{max} = \frac{3}{4} \times \frac{F}{b(h - 2e)},$$

the maximum value of the shear stress at the neutral axis.

Combining the two solutions
The shear stresses acting across a thin skinned sandwich beam are:

at the adhesive interface
$$\tau = \tau_1 \times \frac{F}{2bc}$$

at the neutral axis
$$\tau = \tau_{max} = \frac{3}{4} \times \frac{F}{b \cdot c}$$

as e is negligible compared with h and c.

Conclusion
This shows clearly that the resistance of the sandwich core to shear stress is important and must be tested. It also indicates the importance of the adhesive

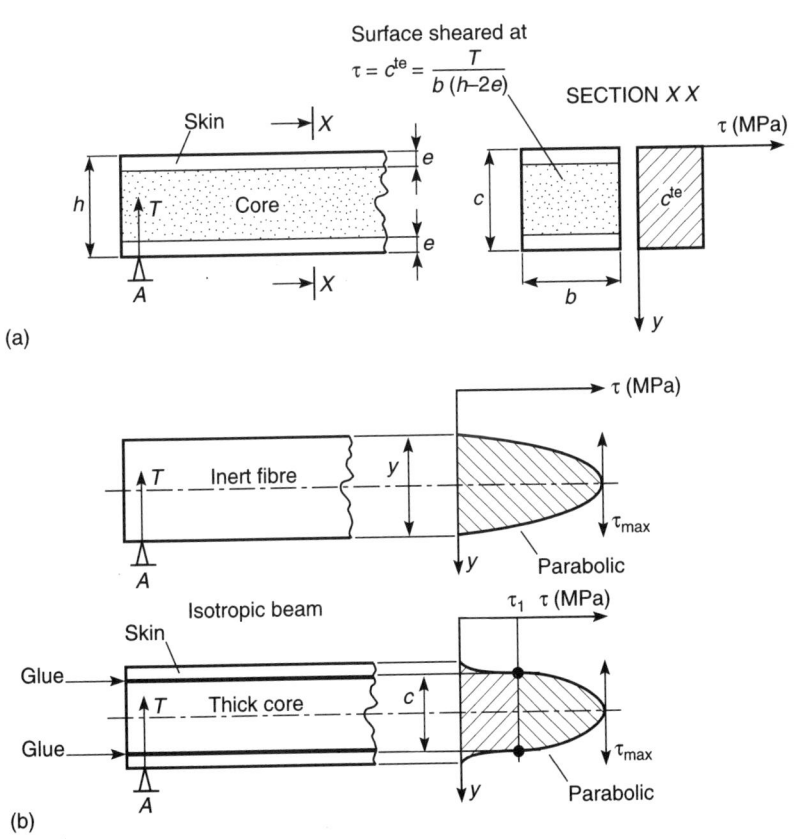

Figure 2.71 Shear stress estimation: **(a)** sandwich beam, elementary method; **(b)** sandwich beam, approached method.

bond between the skins and the core. These calculations are illustrated in Figure 2.71.

(b) More accurate method of calculating shear stress

Daniel Gay's course on the resistance of composite materials (published in France by Hermes) (page 87) gives the values of shear stress τ in a composite sandwich subjected to simple flexure. Let e_p be the thickness of the stratified skin; e_c be the thickness of the core; $h = e_c + 2e_p$ be the thickness of the beam; b the width of the beam $= 1$; E_p and E_c Young's moduli of the skin and core; and G_p and G_c the shear moduli. These are illustrated in Figure 2.72.

Shear stress in the skin–core interface of the sandwich
In the skin at a distance y from the neutral axis, one gets:

$$\tau = \frac{1}{2} \times \frac{\langle GS \rangle}{\langle EI \rangle} \times E_p \left(\frac{h^2}{4} - y^2 \right).$$

The adhesive interface is at a value of $y = e_c/2$; moreover

$$\frac{\langle GS \rangle}{\langle EI \rangle} = 12 \times \frac{G_c e_c + G_p(h - e_c)}{E_c e_c^3 + E_p(h^3 - e_c^3)}.$$

Thus the shear stress within the interface is:

$$\tau_1 = \frac{1}{2} \frac{\langle GS \rangle}{\langle EI \rangle} \times E_p \left(\frac{h^2}{4} - \frac{e_c^2}{4} \right)$$

Maximum shear stress within isotropic core
Assume that the foam core is isotropic. In the foam at distance y:

$$\tau = \frac{1}{2} \frac{\langle GS \rangle}{\langle EI \rangle} \times \left\{ E_c \times \left(\frac{e_c^2}{4} - y^2 \right) + E_p \left(\frac{h^2}{4} - \frac{e_c^2}{4} \right) \right\}.$$

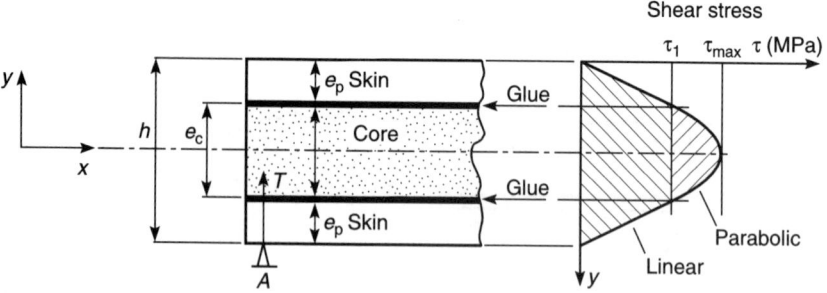

Figure 2.72 Sandwich beam: accurate method (after D. Gay).

The maximum shear stress occurs at the neutral plane where $y = 0$ and is equal to:

$$\tau_{max} = \frac{1}{2} \frac{\langle GS \rangle}{\langle EI \rangle} \times \left\{ E_c \times \frac{e_c^2}{4} + E_p \left(\frac{h^2}{4} - \frac{e_c^2}{4} \right) \right\}.$$

The foam should be tested on delivery to check that its resistance to shear is adequate ($> \tau_{max}$ previously calculated). The above equations require a knowledge of the value of stiffness modulus E_c and shear modulus G_c of the core of the material.

2.5.4 Main test standards

- NFT 56-107 and ISO 845: Determination of the apparent specific gravity of a cellular product.
- NFT 56-106 (no ISO standard): Determination of water and volatile matter content of stiff cellular materials.
- NFT 56-101 (and ISO 866): Compression test for stiff cellular products.
- NFT 56-103 (and ISO 1926): Tensile test for stiff cellular materials.
- NFT 56-104 (no ISO standard): Determination of punch stress of stiff cellular materials.
- NFT 56-118 (and ISO 1922): Shear test for stiff cellular products.
- NFT 56-102 (and ISO 1209): Flexural test for stiff cellular materials.
- NFT 56-117 (and ISO 2440): Accelerated ageing of stiff cellular products.

Note that these standards only concern foams. There are hardly any for the control of the quality of honeycomb materials, as these need to be tested in conjunction with a standard face skin.

2.5.5 Tests on stiff cellular foams

(a) Definition

A stiff foam is a cellular plastic, which after compression that has reduced its thickness by half and sufficient time has elapsed for relaxation (10 minutes), shows a relative dimensional change $r \geqslant 10\%$.

$r \geqslant 10\%$ stiff foam.
$r < 10\%$ the material is a soft foam.

If possible, the test should be carried out on a cube of side $H_0 = 50$ mm, compressed to 25 mm thickness and kept under load for 15 seconds. The residual height H is measured after 10 minutes relaxation. Should the material be less than 50 mm thick, H_0 is taken as the thickness of the recured specimen, as shown in Figure 2.73.

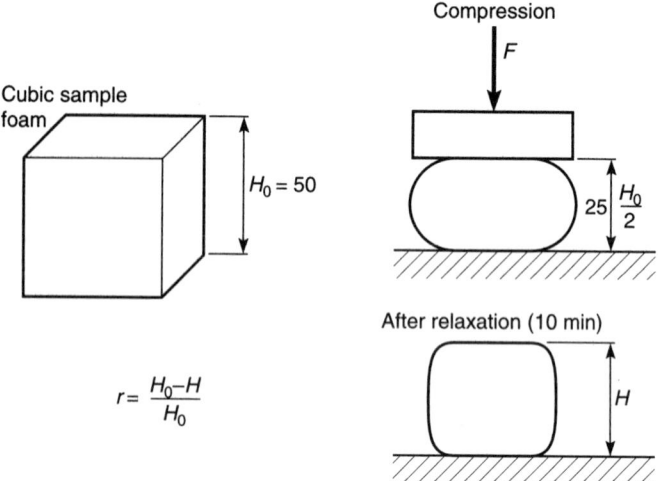

Figure 2.73 Stiffness test.

(b) Determination of specific gravity

This is the simplest test that can be made. Standards NFT 56-107 and ISO 845 will be used.

A minimum of three specimens are needed. They should be rectangular in shape and carefully cut off from as-received foam board. Care must be taken not to crush the structure of the foam when cutting. The volume of the specimen is $V \geqslant 100\,\mathrm{cm}^3$. According to the standards, the specimens must be conditioned for at least 16 hours in the reference atmosphere ($23\,°\mathrm{C} - 50\%$ RH), and then weighed accurately: weight m (g).

Standards: NFT 51-014, ISO 291, EN 62, DIN 50014, BS 2782 part b method 1004.

The volume of the specimen is calculated from the measurements taken: V (mm^3). The apparent specific gravity is ρ:

$$\rho = \frac{m}{V} \times 10^6\,\mathrm{kg/m}^3.$$

Examples: Airex® R62.80, $\rho = 80\,\mathrm{kg/m}^3$; Klegecel® 120, $\rho = 75\,\mathrm{kg/m}^3$.

(c) Water absorption tests

This test is carried out according to NFT 56-106 (no ISO standard). This test may not be needed if the sandwich core is fully dried before use. Drying to eliminate moisture from foams, honeycombs and balsa wood requires 24 hours at $70\,°\mathrm{C}$ maximum in an oven (preferably ventilated).

(d) Further tests for foams

General

Composite skin sandwich constructions are mainly used as flexural beams. When deformed there are three areas of maximum stress.

- A: on the surface where compression stress is at a maximum;
- B: on the surface where tensile stress is at a maximum;
- C: right in the middle of the sandwich core on the neutral axis where the shear stress reaches a maximum.

These are shown in Figure 2.74.

When a sandwich beam is deformed, so is its core, as in Figure 2.75. Within the core:

- at A: there is maximum compression;
- at B: there is maximum tensile;
- at C: there is maximum shear stress.

Adhesion resistance

If the laminated skin–core interface obtained through bonding is perfect, the elastic deformation due to bending of the laminate is transferred to the

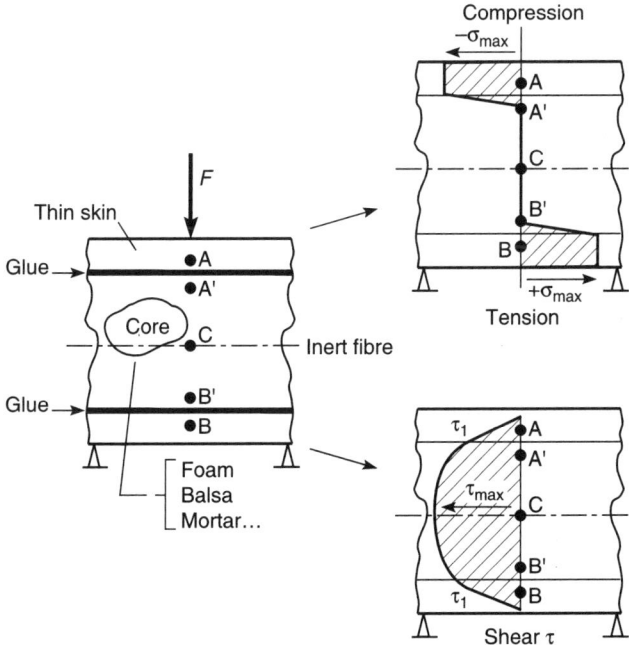

Figure 2.74 Sandwich beam stress distribution.

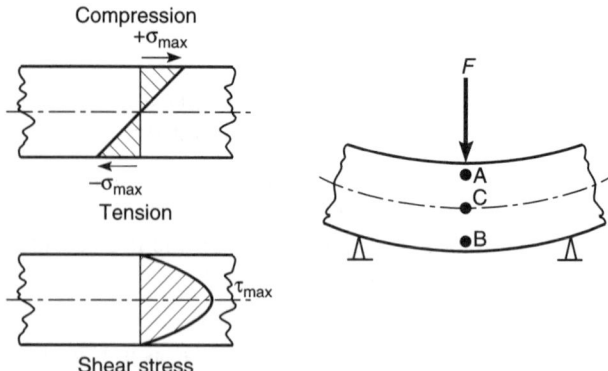

Figure 2.75 Monolithic isotropic beam: stresses.

adhesive. Skins and core are adhesively joined either with resins which may incorporate fillers or with structural adhesives (liquid or film).

The characteristics of resins and adhesives vary considerably according to whether the material has been cold cured (20 °C) or hot cured. The following table gives some values for tensile resistance as a function of the cure.

Table 2.2 Results on tensile test

Resins	R (MPa)	E (MPa)
Polyester UP	30–80	≈ 3000
Phenolic	20–60	≈ 3000
Epoxy EP	40–90	≈ 4000

Note that the strength of resins under compression is four to five times as high as it is under tension. Thus, for a polyester, σ_{max} under compression = 200 MPa. Polymerized adhesive has a low modulus, $E = 3000$ MPa and is easily deformed ($A = 2$–3%) so that it generally adapts to the stress and rarely fails.

Resistance of foam under tension
If the adhesive–cellular foam interface is perfect, the elastic deformation of the adhesive is completely transferred to the foam on bending. Sandwich foams and cores have a very low Young's or flexural modulus and can thus withstand greater strains than the laminate and adhesive ($A > 3\%$). Some values of the modulus E are:

- balsa wood: $E = 2900$ MPa;
- Firet Coremat® : $E = 790$ MPa (infiltrated with resin);

- Airex® R62.80: $E = 27\,\mathrm{MPa}$;
- Klegecel® 120: $E = 44\,\mathrm{MPa}$.

Foams can therefore adapt themselves to stress and will not fail, even though the sandwich skin may be under severe stress as the following example shows.

Consider a sandwich made up of a UD 50/50 glass-epoxy laminated skin and an Airex® R62.80 foam core. Suppose the skin is subjected to tensile stress of 1000 MPa when under flexure. Its failure strength is $\sigma_1 = 1100\,\mathrm{MPa}$. Its stiffness modulus is $E_1 = 30\,000\,\mathrm{MPa}$.

The modulus of the foam E_2 is about 30 MPa, and the modulus ratio is:

$$\frac{E_2}{E_1} = \frac{1}{1000}.$$

In Figure 2.76, foam deformability $\varepsilon\,(\%)$ is assumed to be superior or equal to that of the laminated skin.

The stress the foam is subjected to will be:

$$\sigma_2 = \sigma_1 \times \frac{E_2}{E_1} = 1000 \times \frac{1}{1000} = 1\,\mathrm{MPa}.$$

The strength of the foam under flexural stress is 1.76 MPa, and it will not break.

If the foam cracks, as in Figure 2.77, the local stress σ_2 in the skin is increased up to the failure point and the complete beam may fail.

Should a crack appear in the foam, the deformed beam would fail.

Control of foam under tension
It is required that foams should have a high tensile strength so as to avoid cracking in the tight area (B') of the sandwich beam. This tensile strength can be measured upon delivery of the foam.

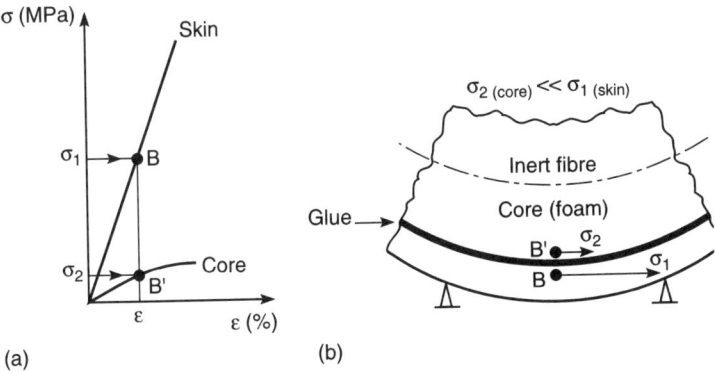

Figure 2.76 Foam resistance under flexure: **(a)** traction curves; **(b)** beam under flexure.

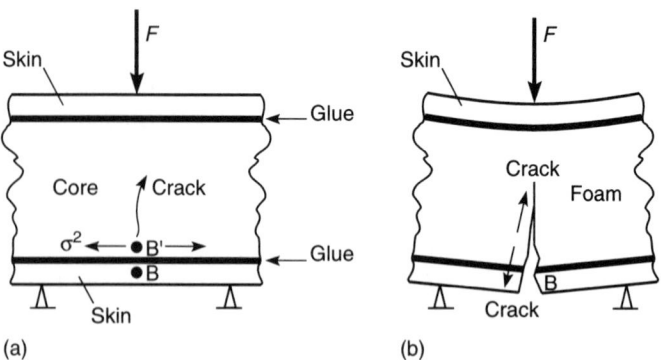

Figure 2.77 Sandwich failure: **(a)** initiation; **(b)** development.

Some flexural strengths are:

- Balsa: $\sigma_{max} = 8\,\text{MPa}$ (excellent).
- Firet Coremat®: $\sigma_{max} = 6.5\,\text{MPa}$ (resin infiltrated).
- Airex® foam: $\sigma_{max} = 1.8\,\text{MPa}$.
- Klegecel® 120 foam: $\sigma_{max} = 1.5\,\text{MPa}$.

Compression strength of foam
If the yield strength of the foam is exceeded under compression, the cells will be crushed, and this may lead to subsidence, in a floor panel, or buckling of a column or panel, as illustrated in Figure 2.78.

Examples: Klegecel® 120: $\sigma_c = 1.10\,\text{MPa}$; Airex® R62.80: $\sigma_c = 0.85\,\text{MPa}$.

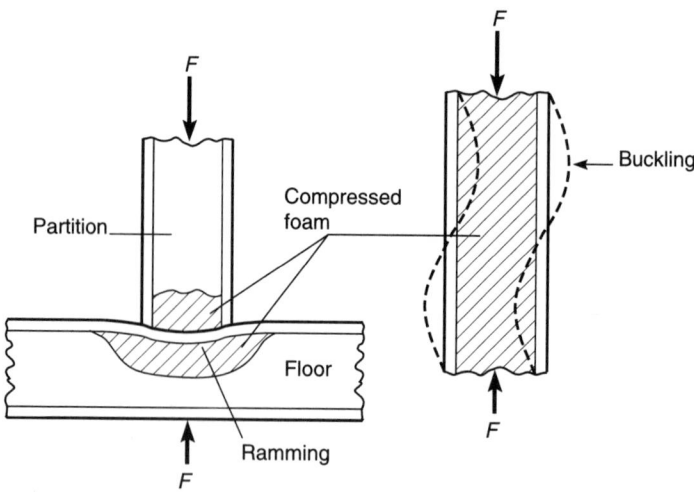

Figure 2.78 Compressed sandwiches.

Shear strength of foam

Lastly, it is necessary to check shear resistance.

Examples: Klegecel® 120: $\tau_{max} = 0.65$ MPa; Airex® R62.80: $\tau_{max} = 1.20$ MPa.

Tensile test

The appropriate standards are NFT 56-103 and ISO 1926, and a minimum of five specimens are needed. These must be cut carefully from the foam (thickness $= e$), and aluminium reinforcements glued to both ends of the specimen. Specimens are conditioned for 24 hours at 23 °C and 50% RH and then stressed slowly (rate < 5 mm/min) until they break when the load reaches F_r. The machine used for the test must be fitted with special jaws as traction is exerted on the reinforcements through axles of diameter $\phi = 30$ mm, as shown in Figure 2.79.

The cross-section of the specimen is $S = 25 \times e$ (mm²); the ultimate strength is $\sigma_{max} = F_r/S$ (MPa). The tensile stress curve is gross elastic linear: the foam's stiffness modulus is

$$E = \tan\alpha = \frac{\sigma}{\varepsilon_r} \text{ (MPa)}$$

Compression test

The test standards are NFT 56-101 and ISO 844. The test continues until crushing reaches 10%. The specimens are conditioned for 16 hours at 23 °C and 50% RH.

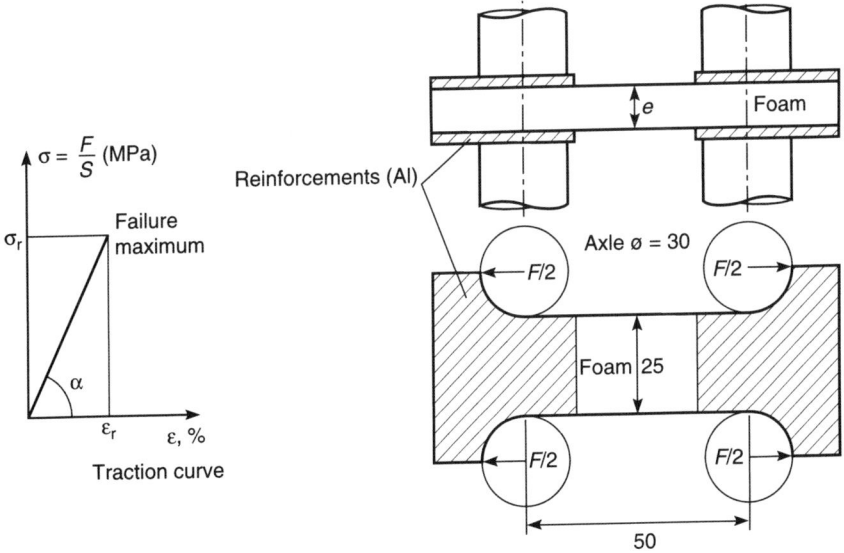

Figure 2.79 Tensile test on foam.

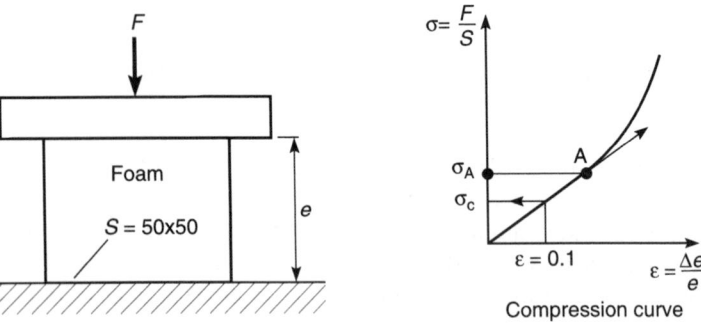

Figure 2.80 Compression test on foam.

Five cubic specimens measuring $50 \times 50 \times 50$ mm or $50 \times 50 \times e$ (thickness of the foam board) are subjected to compression at slow speed, as shown in Figure 2.80.

The characteristic which is measured is the compression stress σ_c obtained when the strain $\varepsilon = \Delta e/e$ reaches 10%:

$$\sigma_c (Pa) = \frac{F_c \ (N)}{S \ (m^2)}$$

The compression stress σ_A which corresponds to the end of the elastic linear range of the foam can also be calculated.

Shear test

Standards ISO 1922 and NFT 56-118 for determination of the apparent shear modulus G require a lengthy and costly test. Five specimens are cut and glued on to thick metal sheets with an epoxy adhesive which has a very high resistance to shearing, as in Figure 2.81. After being conditioned at 23 °C and 50% RH, the specimens are subjected to tension at a very low speed (rate < 1 mm/min) until rupture occurs within the foam.

The load elongation curve of the mobile jaw relative to the fixed one is recorded. This curve (F, Δ) has a linear part, the slope of which is β. The relation

$$G = \frac{e}{lb} \times \tan \beta \ (MPa).$$

yields the shear modulus G Coulomb's modulus.

This test can also give an assessment of maximum strength τ_{max}, of the stiff foam:

$$\tau_{max} \simeq \frac{F_m}{bb} \ (MPa).$$

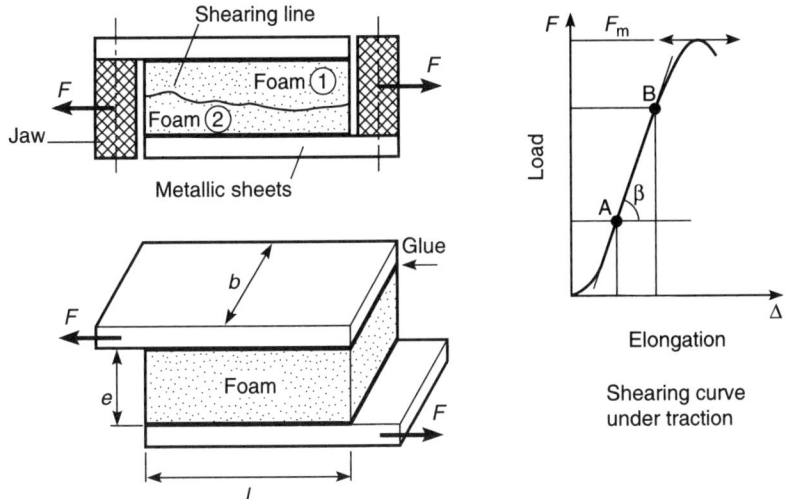

Figure 2.81 Shearing test on foam.

Other tests may also be used, for example,

- 'rail shear test' – ASTM D4255;
- Iosipescu test.

The rail shear test, which was devised for monolithic composites, uses a sheet of the material clamped along its edges as in Figure 2.82. There are two

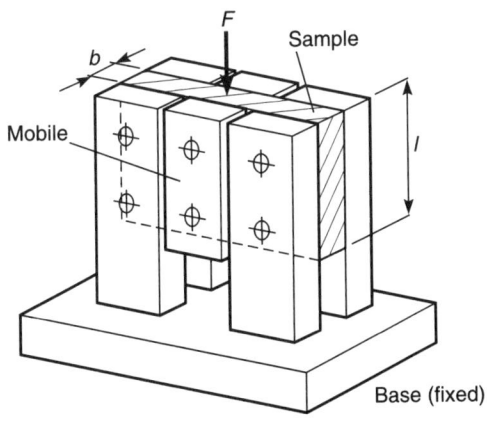

Figure 2.82

shear sections: $S = 2\,lb$. The curve (F, Δ) is recorded, and the readings are analysed in the same way as those obtained during other shear tests (NFT 56-118 or ISO 1922).

The test can also be carried out on a foam specimen which is glued to the fixed, or moving, parts of the test machine.

2.5.6 Tests on balsa wood

Balsa wood is very light ($\rho = 40\text{--}200\,\text{kg/m}^3$) and is always used with the wood fibres perpendicular to the laminated skins, as shown in Figure 2.83. Balsa wood is made up of parallel cells (99% voids) so that its cross-section looks like that of a honeycomb. Various tests should be carried out on the as-received material including:

- specific gravity (on the complete board);
- absorbed moisture ratio;
- tensile strength perpendicular to the fibres;
- compression parallel to the fibres;
- shear parallel to the fibres.

The same test and standards recommended for stiff foams, as previously described, can be used.

Note that balsa wood is apt to absorb moisture. It must therefore always be stoved carefully and it must be checked that the resin used to bond on the skins can penetrate into the wood to give a good bond. The balsa wood used must, however, retain 12% moisture.

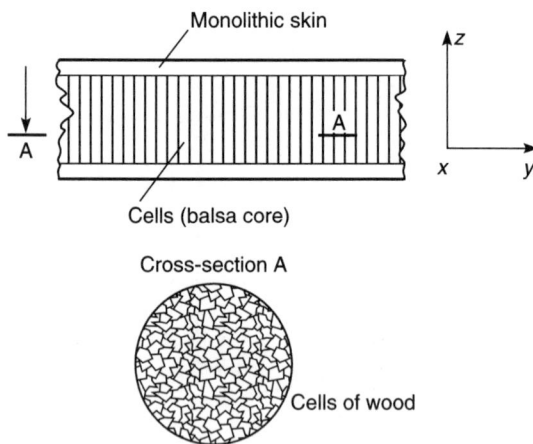

Figure 2.83 Structure of sandwich.

For a Belco® balsa standard, we find:

- specific gravity $\rho = 128\,\text{kg/m}^3$;
- compression strength $\sigma_c = 9.5\,\text{MPa}$;
- tensile strength $\sigma_t = 12.7\,\text{MPa}$;
- shear strength $\tau_{yz} = \tau_{xz} = 1.8\,\text{MPa}$.

2.5.7 Tests on honeycombs

Standards do not call for tests on as-delivered honeycombs, though they do recommend testing the adhesion between laminated skin and the honeycomb. However, honeycomb suppliers such as Ciba-Geigy (who sell Aerolam-Aeroweb®), Induplast, who sell polypropylene honeycomb covered with non-woven glass mat, and other companies, guarantee certain properties, namely:

- specific gravity;
- shear strength τ_{yz} and τ_{xz};
- compression strength.

It is possible to test the material for these properties.

(a) Specific gravity test

Standards NFT 56-107 and ISO 845 are used (as for stiff foams and balsa wood). The method consists in measuring accurately the volume V of honeycomb (including the cells), as in Figure 2.84:

$$V = lbe, \quad \rho = M/V\,(\text{kg/m}^3).$$

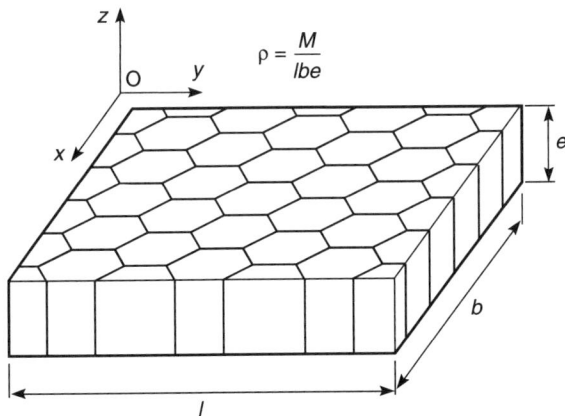

Figure 2.84 Honeycomb specific gravity.

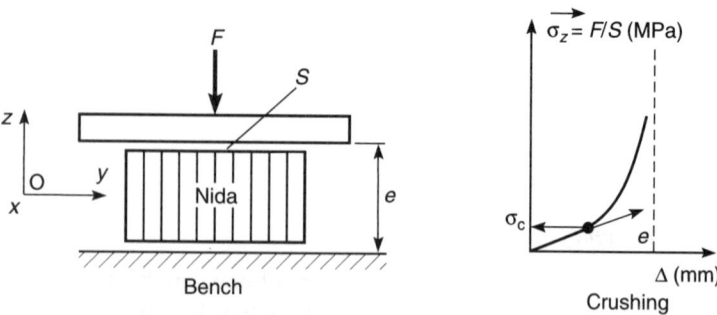

Figure 2.85 Compression test.

Honeycomb specific gravity is

$$\rho = \frac{M}{lbe}.$$

Examples are Nida Aeroweb® core in 3003: $\rho = 29\text{--}83\,\text{kg/m}^3$; Nida Nomex (Du Pont de Nemours): $\rho = 64\,\text{kg/m}^3$.

(b) Resistance to compression

This is determined according to standards NFT 56-101 and ISO 844 (see tests carried out on stiff foams). This is illustrated in Figure 2.85.

Example: $\sigma_z = 4\,\text{MPa}$ (Nida Aeroweb).

Note that this test can also be carried out on the sandwich, i.e. the honeycomb plus laminated skins. Ciba-Geigy's Aerolam® boards can also be tested in the same way.

(c) Resistance to shearing

Stiff foams can be considered to be isotropic materials, unlike honeycombs and balsa which possess different properties depending on how stress is applied to the individual cells. Honeycombs are made as shown in Figure 2.86: sheets are bonded together with glue and then expanded.

Shearing principally occurs along planes xOz and yOz. Resistance to shearing on plane xOz is τ_{xy} and τ_{zy}. Resistance to shearing on plane yOz is τ_{yx} and τ_{zx}.

The values we get for such resistances are different (unlike the case with balsa wood). We have:

$$\tau_{\text{plane } yOz} > \tau_{\text{plane } xOz}$$

or

$$\tau_{zx} > \tau_{zy}$$

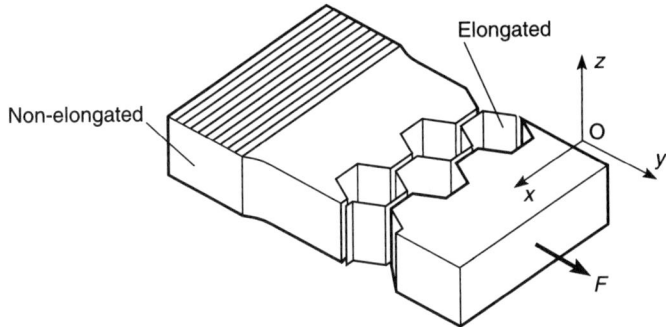

Figure 2.86 Honeycomb expansion.

and

$$\tau_{yx} > \tau_{xy}$$

which is understandable for the yOz plane intersects the areas where cells are glued to one another so that the thickness is doubled, as shown in Figure 2.87.

Note that the values of τ_{zx} and τ_{zy} depend on the thickness e of the layers of which the honeycomb is made, on the way these layers have been bonded,

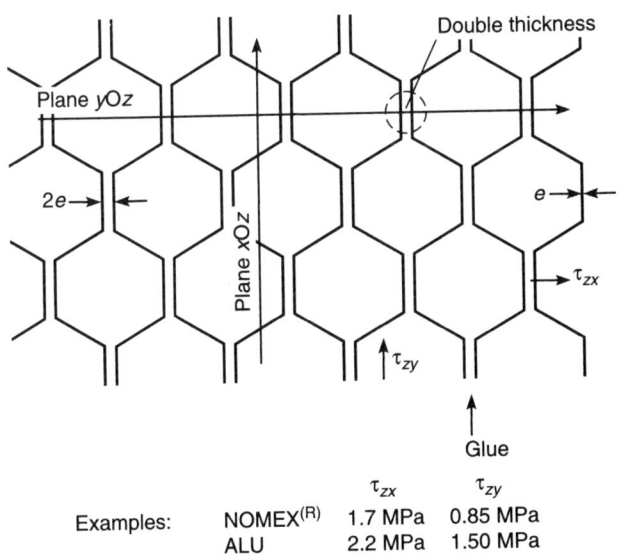

Examples:		τ_{zx}	τ_{zy}
	NOMEX$^{(R)}$	1.7 MPa	0.85 MPa
	ALU	2.2 MPa	1.50 MPa

Figure 2.87 Shearing of honeycomb.

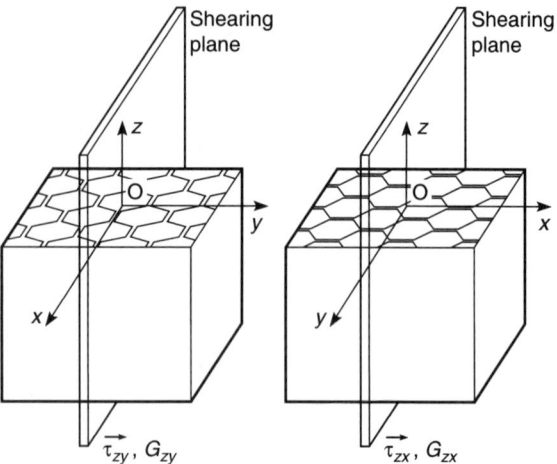

Figure 2.88

and on the dimensions of the cells (given by the diameter of the circle inscribed in a hexagonal cell).

In lieu of a better standard, the ASTM 'rail shear test' can be used (Section 2.5.5). The honeycomb is bonded with epoxy resin to the fixed and moveable part of the testing machine. According to the orientation of the cells, we will have the arrangement shown in Figure 2.88. This test is costly, because of the time required to prepare the specimens and set up the testing equipment.

2.5.8 Testing Coremat®

(a) Definition

This type of sandwich core is manufactured by Firet (The Netherlands). It consists of a non-woven mat of polyester fibres and expanded microballs of polyvinylidenechloride (PVDC) bonded together by a polyacrylic adhesive, as shown in Figure 2.89.

The fibre mat (a few cm in length) keeps the balls from moving. The balls make up about 50% of the volume. Firet Coremat® comes in 50 m rolls (similar to glass mat). The thickness can vary between 1 and 5 mm. It absorbs 0.5 litre of resin per m² per mm of thickness.

(b) Characteristics

The properties of the impregnated and cured material are:

- density $\rho = 640 \, \text{kg/dm}^3$;
- tensile strength $\sigma_m = 3.8 \, \text{MPa}$;

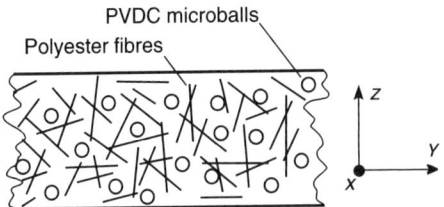

Figure 2.89 Coremat® structure.

- flexural strength $\sigma_t = 6.6\,\mathrm{MPa}$;
- shear strength $\tau = 1.10\,\mathrm{MPa}$;
- compression strength $\sigma_e = 22\,\mathrm{MPa}$.

(c) Tests

Tests upon delivery are for:

- thickness;
- dimensions;
- moisture content.

Note that it is as well to dry out the material before use. In this case there is no need to measure the moisture content of as-received material.

Tests after resin impregnation and polymerization by the user (Coremat multilayered laminate) are for:

- specific gravity;
- strengh properties (see Section 2.5.8(b)).

2.5.9 Testing microballs

(a) A definition

Glass microballs are mixed with resin to produce mortars (with solid microballs) and syntactic foams (with hollow microballs). Only hollow microballs are used to produce lightweight sandwich cores. Hollow microballs are made of borosilicate glass and have a wall thickness of 0.5–3 microns. The bubbles contain an inert gas (N_2), and are sized to improve adherence to the resin. They vary in size (diameter) from 5–200 microns. Their resistance to pressure varies from 30–250 bars.

(b) Tests

The characteristics to be tested are the following:

- apparent specific gravity;
- average diameter;

- moisture content ratio;
- resistance to pressure.

Testing for density
After the microballs have been dried by heating, they are put into a container (volume V_0, mass m_0) which is then weighed: mass m_1. The apparent density is

$$d = \frac{m_1 - m_0}{V}.$$

Depending on the type of microballs, the density varies from 0.08 to 0.23 g/cm^3. The American standard ASTM D 3101-72 gives a description of the test.

Resistance to crushing
The hydrostatic crushing test is carried out according to ASTM D 3102-72. The microballs are placed into a nacelle which is immersed in a high pressure chamber. When they are taken out of the chamber, the microballs are examined with a scanning electron beam microscope to assess damage.

Average diameter
The average diameter of the microballs can be verified with an optical or a scanning electron microscope. Because of the variation in sizes a mean diameter is determined.

2.6 ADHESIVE CONTROLS

2.6.1 Need for control

The adhesives used are raw materials which must be controlled upon delivery and just before use. The adhesives are organic resins and may thus deteriorate with time. Those adhesives which harden when heated are especially susceptible to ageing during storage.

A strict control of storage conditions must be ensured. The adhesives must always be stored in a dry place. According to their nature they must be stored at an ambient temperature, or around 0 °C, or around −20 °C.

Suppliers should give details concerning storage conditions as well as the lifetime to be expected under these conditions.

An adhesive stored in a freezer will deteriorate rapidly at room temperature, having a life of a few days or less.

2.6.2 Adhesives used

For bonding the skins to a sandwich core, various resins are used. The type depends on the service temperature of the artefact.

(a) Types

These resins are in one of the following forms:

- prepared from a liquid or paste in one or two components which are mixed just before use; the mixture (resin + hardener) is thinly spread, with a brush or a notched scraper, on to all surfaces to be bonded
- or as a thin film or film supported on a fine mesh.

This product is widely used for structural bonding. The film of resin, which is sandwiched between two non-adhesive sheets of siliconed paper, comes in the form of a roll, similar to a roll of prepreg. The resin must be kept in a freezer.

Some films contain, mixed in with the resin, glass microballs to limit the thickness of the joint line. Others are carried on a fine woven or non-woven glass fibre or synthetic yarn to help with manipulation in use.

(b) Principal groups (Standard T76-011 (1980))

1. Phenol – formaldehyde vinyl.
2. Epoxies (the most commonly used). These are easy to use and adhere to most materials providing good resistance to shearing and peeling, but they may age and their ability to resist high temperatures is related to the curing temperature.
3. Phenolic epoxy – good control of humid ageing.
4. Nitrile epoxy – cross-linking at temperatures lower than the above.
5. Phenolic nitrile – good performance up to 120 °C and good resistance to peeling. Unsuitable for bonding honeycomb surfaces.
6. Nylon epoxy – excellent resistance to the surrounding temperature, but poor response to heat and humid ageing.
7. Polyaromatic polyimid – the most resistant to heat from 200 °C to 300 °C. Unsuitable for bonding honeycomb surfaces.

2.6.3 Physico-chemical control

The quality control of adhesives is similar to that of epoxy resins. We can thus refer to Section 2.2. The physico-chemical methods which detect ageing of adhesives are very important. A fresh resin is used for comparative purposes. Basically let us cite the following test procedures.

Gel time at ambient or high temperature (depending on the use/type of the resin) is established by using a curing simulator, or, better still measuring viscosity (cf. Section 2.2.1).

The differential scanner calorimeter (DSC) is the preferred tool to determine global reactivity though the degree of reactivity and its change with time can be followed with the Brookfield viscometer: standard AFNOR T76-116.

The DSC allows the fast, efficient study of all types of adhesives: liquid, paste or film. Moreover, the adhesive may be polymerized at its real temperature of use or undergo the curing cycle suited or planned for the piece.

Aged adhesives show a smaller cross-linking reaction energy ΔH than does fresh material.

2.6.4 Mechanical testing

The measurement of the shear characteristics are the best, mechanical, way of evaluating the properties of an adhesive layer, though great care must be taken to prepare identical test pieces. The results show considerable variation and do not give as convincing a picture of ageing as does the DSC technique.

(a) Shear strength under tensile loading

The adhesive layer is formed between two sheets of ASTM 5000 aluminium alloy. This must be degreased with solvent before treatment in a hot alkaline solution, followed by pickling in a chrome sulphate bath. These treatments eliminate the grease on the metal surface and the layer of Al_2O_3 present always on the surface.

A fresh chromium anodization solution (less than eight hours old) is recommended in order to form a microporous surface on the metal and to eliminate the risk of adhesive failure at the metal–adhesive interface. For a valid test result cohesive failure must occur through shearing in the adhesive itself (as shown in Figure 2.90).

The standards applicable are:

- AFNOR T76-107 (1979)
- ASTM D 1002-72 (1978)
- ASTM D 3258-76 (1981)
- AFNOR T76-141 (1988)
- ASTM D 3165-73 (1979).

The shear test piece is composed of two ribbons of stamped sheet metal,

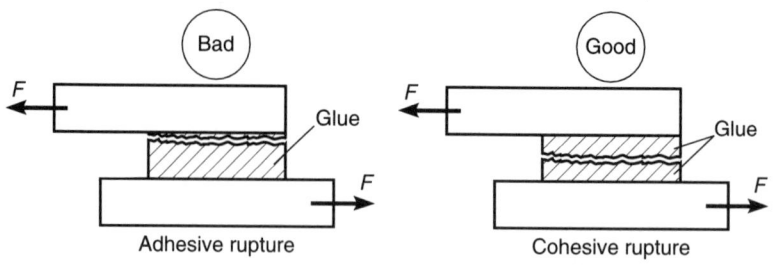

Figure 2.90

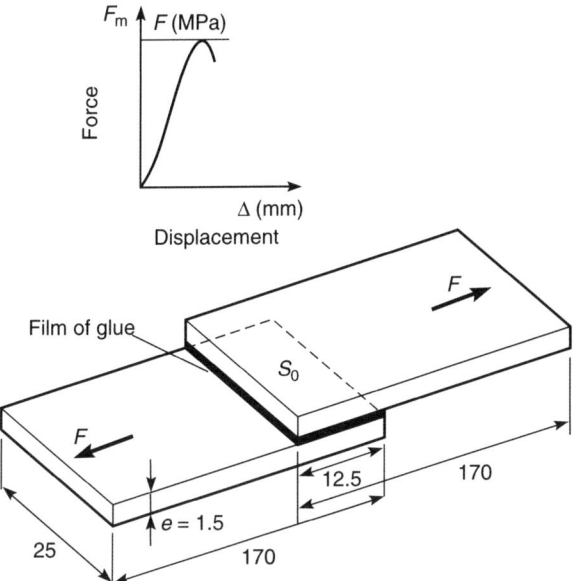

Figure 2.91 Standard shearing–tensile test.

without flash, as shown in Figure 2.91. The adhesive coating is very thin, 0.1 to 0.2 mm, and may be a piece of film adhesive. Rubber gloves should be worn to avoid touching either the metal surface or the adhesive.

Curing is carried out at the pressure, temperature and for the time recommended by the manufacturer; for example for an epoxy adhesive 3 bars (in an autoclave) at 120 °C for eight hours.

The stress/strain curve gives the maximum force F_m at the failure of the adhesive surface, S_0. The shearing strength of the adhesive is:

$$\tau = \frac{F_m}{S_0} \text{ (MPa)}.$$

Aged adhesives have a lower shear strength.

(b) Thick adherent shear test (TAST)

There is a tendency for the metal pieces to self-align during stressing causing normal forces to be generated at the tips of the adhesive–metal interface, as in Figure 2.92. To avoid this a thick, rigid, test piece is used. This is the standardized ASTM test piece for the thick adherent shear test (TAST), ASTM D 32-58-76.

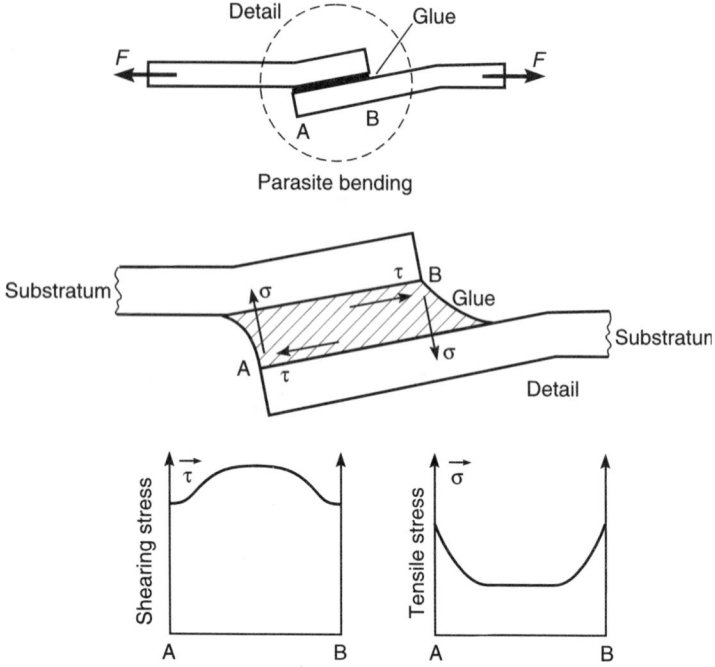

Figure 2.92 Thick adherent shear test.

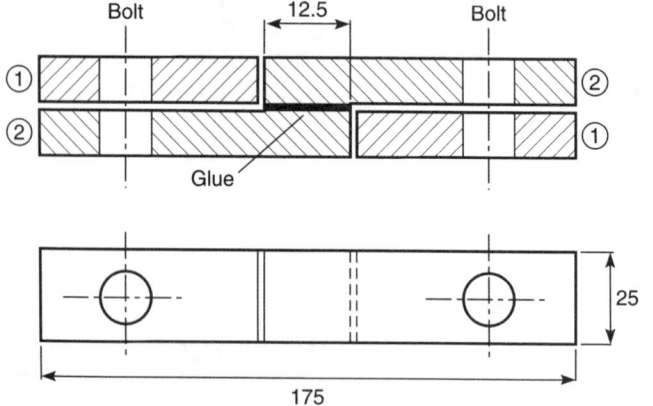

Figure 2.93 TAST test piece.

This system allows the joint to be stressed without bending and hence failure in the adhesive to occur due to shear stresses. Each test piece is made up of two parts. Part 1 can be reused.

At least five tests must be carried out to give a mean and standard deviation (see Figure 2.93).

Mould controls | 3

3.1 MOULD TYPES

Moulds may be made from various materials, depending on the moulding process, the component to be made and the number required.

Possible materials range from the simplest (plaster) to the most expensive (invar (Fe-Ni36)).

3.1.1 Recommendations

The more artefacts that are required, the more important it is to choose a rigid material which will not wear and not be influenced by atmospheric conditions (humidity and temperature). If high mould pressures are needed a strong, rigid mould material is called for.

The thinner, and so more precise, the piece to be made, the more important it is to choose a low shrinkage material from which to make the mould.

3.1.2 Production, nature and capacity

Plaster moulds are used in polyester contact moulding. They can be used to make a maximum of five replicates.

Silicon rubber moulds are used in polyester contact moulding or vacuum bag moulding with fabrics that have been preimpregnated with epoxy resin. They can produce one to ten components.

Wood moulds are used in contact moulding or projection. Depending on the type of wood used and its preparation, they can produce 10–50 components.

Thermoset laminated composite moulds are used in contact moulding; moulding by projection; bag moulding (vacuum of autoclave); moulding assisted by a vacuum or low-pressure injection; and cold compression moulding.

These mould types are common and not very costly. Depending on the structure of the mould adjacent to the laminated skin, they can produce 50–100 artefacts.

Concrete resin moulds (gravel + resin + steel armatures) are used for vacuum injection moulding between two moulds; low-pressure (3 bars) injected moulding between two moulds; and cold compression moulding.

These moulds are capable of producing a large number of replicates: up to 1500 pieces with concrete polyester resin, and up to 5000 pieces with concrete epoxy resin covered with a metal shell. The shell is made of a film of silver and a layer of a few millimetres of nickel deposited by electrolysis.

Metal moulds are used for all types of moulding, but usually for large volume production of parts needing high dimensional precision; also for very large components.

Some examples are:

- A very large component (e.g. a mine sweeper): length 60 metres; number 30 units; mould in polished, soldered sheet steel.
- A piece of high precision: e.g. a satellite telecommunication dish: length 3–5 metres; number 6 units; invar mould, polished.
- A large quantity of accurate parts: e.g. helicopter blades: length 9 metres; number over 2000 units; mould of treated and polished aluminium (2024) ASTM.

The main materials used in the making of metal moulds are:

- soldered sheet steel;
- aluminium alloys: 5000 or 2000 ASTM;
- heat treated and chromium plated steel (this will be good for 500 000 components!);
- invar, Fe-Ni36, because of its near zero expansion coefficient; all metal moulds are carefully polished;
- ceramic mould or pulforming die.

3.2 CONTROL PROCESS

A mould is expected to be dimensionally stable at all times; to be rigid and resistant during the clamping and closing period; to withstand shock and thermal fatigue; and to ensure demoulding without difficulty.

As a result, the controls on moulds include:

- dimensional
- surface condition
- heat cycle and heat mechanism when integrated in the mould.

3.2.1 Dimensional control

The dimensions of the mould must be as required to produce a part with the correct tolerances and not affected by pressure, temperature or time (cf.

Section 5.3). To check the dimensional stability of the mould, the following checks must be carried out:

1. cold, after the making of the mould;
2. hot, under the moulding conditions;
3. after one or more thermal cycles simulating the use of the mould including its return to ambient conditions. A comparison will finally be made between 1 and 3.

It is important to control the drafts which must be at least 5% to ensure an easy demoulding.

3.2.2 Surface condition control

The mould's surface must be smooth and thus polished. The condition of the surface is checked in accordance with the customary metrology techniques (cf. Section 5.3). The mould must have not sharp edges.

3.2.3 Mould heating control

The homogeneity of the mould's temperature is extremely important for successful curing (cf. Chapter 4.2). Where heated moulds are used, one of the following should be employed:

- metal with hot oil circulation and cooling circuits;
- composites with laminated boards which include heating elements or heating fabric made of carbon.

It is necessary to carry out heating tests and check the uniformity of the temperature with the aid of thermocouples.

4 Operating control in manufacturing

4.1 LAMINATING CONTROL

4.1.1 Main processes used

The manufacturing processes for thermosetting laminates are numerous:

- contact moulding;
- spray-up moulding;
- bag vacuum moulding, or pressure bag moulding;
- high- or low-pressure injection moulding;
- cold compression moulding;
- rotational moulding;
- filament winding;
- pulforming;
- production of sandwich boards.

Some of this technology can be automated. This chapter will not deal with all types of moulding technology but will cover some examples in order to deduce the principles of manufacturing growths:

- contact moulding by a 'wet process';
- vacuum bag moulding, also called 'pressure bag moulding';
- sandwich moulding;
- filament winding.

4.1.2 Contact moulding by wet lay-up

This is the most common fabrication process since it needs little investment and is adapted to one-off products or small numbers. It is used by a wide range of composite fabricators.

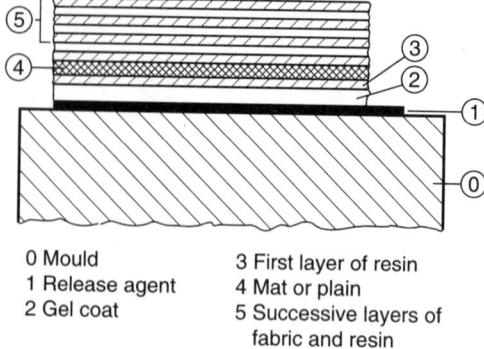

0 Mould   3 First layer of resin
1 Release agent   4 Mat or plain
2 Gel coat   5 Successive layers of
                             fabric and resin

Figure 4.1 Contact moulding.

(a) Principle

Each layer of fabric is thoroughly saturated with resin as it is laid in or on the mould. A brush or roller is used to ensure penetration of the layers. Usually glass fabric is wetted with a liquid UP polyester resin. The design in Figure 4.1 shows how the different layers are placed starting from the mould surface.

(b) Controls to be carried out in chronological order

1. Check the cleanness of the mould, particularly the absence of dust.
2. Apply a controlled amount of a release agent. A coloured release agent is useful.
3. Control the deposit of the gel coat for uniformity and mass (300–400 g/m^2) by weighing the charge of material in the applicator before and after use.
4. Check that the coating has practically gelled (is only just sticky) before proceeding.
5. Check that the resin mix has been properly prepared (correct amounts of resin, accelerator, catalyst).
6. Weigh out the quantity of resin to be used for each layer bearing in mind the nature of the surface to be impregnated, the mass of fabric in the layer and the amount of resin needed in the laminate.
7. Check that the fabric is dry or has been properly stoved before use.
8. Check the mass per unit area of the fabric in grams by weighing a small sample. This will enable the mass in a layer to be calculated and the overall fabric content of the finished composite determined (cf. para 6).
9. Do not forget to spread the first layer of resin, a slightly sticky gel coat.

10. Lamination is carried out layer by layer. The design office should indicate workshop instructions on the lay-up plan, i.e.
    - the nature of the surface layer and its mass per unit area;
    - the number of layers of fabric;
    - the nature of the fabrics – kind of weave and mass per unit area;
    - the positioning of each fabric layer with respect to reference points on the mould;
    - the width of the strips and the position of joints, with covers;
    - the stacking sequence.
11. Check for the absence of air bubbles in each layer.

All these points are to be checked as the laminating takes place. Obviously it is impossible and unacceptable for a supervisor to be standing over every laminate worker. The golden rule in composite workshops is thus to *trust the laminate worker* and leave him/her to his/herself to complete the job properly. This calls for full professional training.

Remember: *control is no longer possible after lay-up*.

One technique consists of precutting all the fabric layers in advance and numbering them before stacking them in the opposite order of use. This stack, already carefully controlled, is then given to the laminate worker.

Another technique consists of leaving the non-soaked edges of the fabric outside the mould. The polymerization control follows the laminating operation (cf. Section 4.2).

4.1.3 Vacuum bag moulding and pressure bag moulding

These techniques allow one to obtain a small number or a series of good-quality components having a complex form.

(a) Principle

A flexible liner or sheet is laid on the laminate, inside the mould.

The compaction of the layers of material is ensured either by an external pressure of several bars in an autoclave, or by atmospheric pressure when the bag is evacuated ($\rho = 0.5$–0.8 bar).

As this technique is applied to the production of quality components, it is often used with prepregs and an epoxy die so as to obtain the maximum percentage of strengthened fibres in the required directions and the minimum porosity.

Figure 4.2 shows the placing of material during a preimpregnated moulding (i.e. dry process).

(b) Controls to be carried out in chronological order

A preliminary DSC control of the prepreg gives a check on its age and curing characteristics.

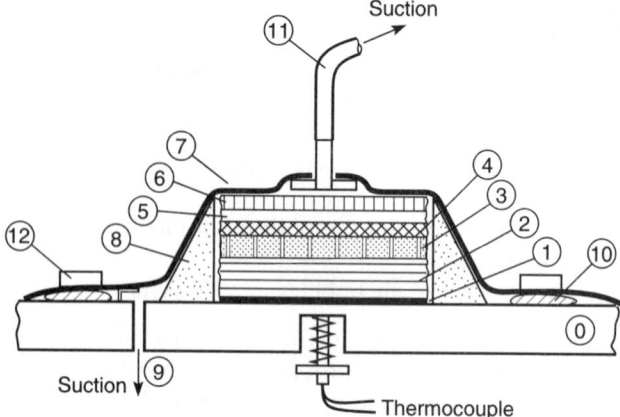

0 Mould (steel, aluminium, composite)
1 Release agent (PTFE teflon)
2 Layers of prepreg
3 Perforated separation layer (bleeder lease)
4 Absorption fabric (not necessary for viscous resin)
5 Non-perforated terfane
6 Breather (air draining fabric)
7 Water-tight bladder (polyamid, silicone rubber)
8 Wedge (silicone or felt resin absorber)
9 Vacuum rod (optional)
10 Sealer
11 Vacuum pipe
12 Clamping device

This diagram is necessary for the control
of work and stacking.

Figure 4.2 Vacuum bag moulding.

1. Check the cleanness of the mould.
2. Control the application of the release agent or release film.
3. Stack the preimpregnated sheets layer by layer taking into account their orientation in relation to the mould ($\pm$ a degree or two). This orientation indicated on the lay-up plan must be adhered to.
4. Control the number of layers including protective films left after laminating. *A protective film must not be left in position in the laminating process.* To avoid this (irreparable) mistake the films are coloured.

   To control the laminate, a similar sequence can be laid up next to the mould on a flat surface at the same time, as in Figure 4.3. The number of layers and their (correct) positioning can be checked.
5. Unidirectional fabrics (UD) have a limited width. To produce large pieces,

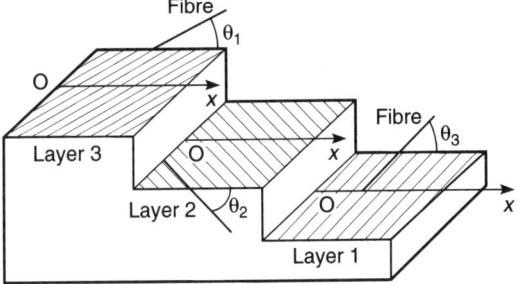

Figure 4.3

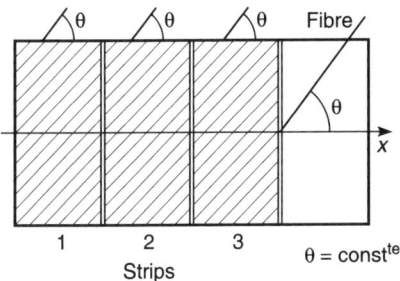

Figure 4.4

strips of UD fabric must be laid side by side as show in Figure 4.4. In this case it is very important that, in each layer, the fibres butt up against one another at a join so that the layer behaves as if it were continuous.

6. Check the setting up of the perforated release film (the role of which is to let through excess resin so that it can drain off); of the bleeder fabric (whose job is to soak up the excess resin); of the resin-tight (non-perforated) film used to stop the resin from blocking the airhole; and of the breather cloth (glass or cotton).

7. Check the positioning of the vacuum outlets around the equipment and at a distance L from the stack of prepreg fabric. In this case the vacuum is created by the rods (autoclave moulding with slight vacuum) as in Figure 4.5. Generally for bag moulding, valves are used.

8. Check that a piece of glass woven fabric has been placed under the vacuum valves in the bag to improve the distribution of pressure, as in Figure 4.6.

9. Check the continuity of the resin tight sealers.

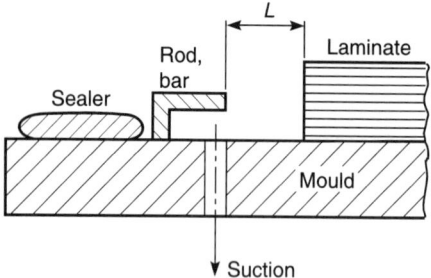

Figure 4.5

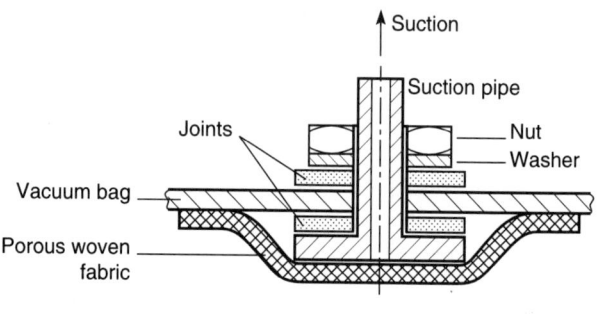

Figure 4.6

All the above operations are carried out by the laminate worker who alone controls the whole process.

Follows: (cf. Section)

- Control of vacuum or autoclave pressure (cf. Section 4.2.1),
- Control of curing (cf. Section 4.2.1).

4.1.4 Sandwich board moulding

There are various processes used to make sandwich boards. We are interested in those techniques which result in shaped pieces, small quantities or one-offs, co-curing the two skins and bonding with the sandwich core in one operation. The process requires a vacuum bag, as in the previous vacuum bag moulding route, and sometimes an autoclave.

We will now look at examples of wet and dry moulding.

(a) Contact moulding – wet process

The operating method is shown in Figure 4.7.

The mould can be of any shape; in this case the core of the sandwich is divided up into blocks which can be linked by a resin-mortar and glass microballs, as shown in Figure 4.8.

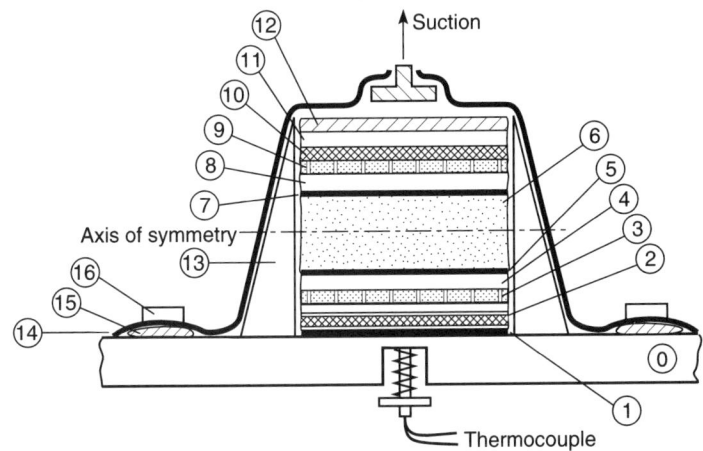

0 Mould (steel, alu, composite)
1 Release agent (wax)
2 Bleeder to soak up excess
  of resin (felt)
3 Porous release film
4 Gel coat (long gel time)
5 Wet laminating of lower skin
6 Sandwich core
7 Wet laminating of upper skin
8 Gel coat
9 Porous release film
10 Bleeder to soak up excess
  of resin (felt)
11 Water-tight release film
12 Breather
13 Silicone wedge
14 Vacuum bag
15 Water-tight sealer
16 Clamping device

Figure 4.7 Wet moulding of sandwich board.

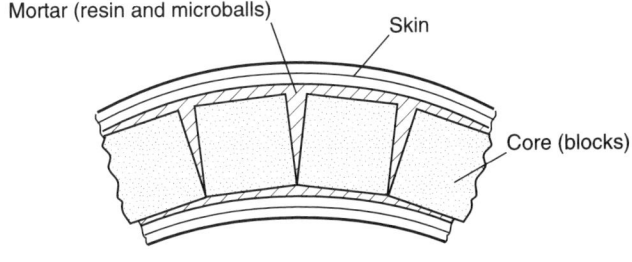

Figure 4.8

Controls to be carried out chronologically by the operator
These are similar to those required for vacuum bag (cf. Section 4.1.3) and contact moulding (cf. Section 4.1.2):

- cleanness of the mould;
- spreading of release agent;
- positioning of bleeder cloth (2) and porous release film (3);
- uniformity and thickness of gel coat;
- waiting for the gel coat to become tacky before laminating;
- checking the preparation and amount of resin;
- checking the fabric for dryness;
- controlling the fabric weight as per the lay-up plan;
- putting a fine layer of resin on the partially cured gel coat;
- then laminating layer by layer, checking the number (don't forget!), the orientation of the layers (cf. plan) and the position of joints;
- checking that there are no air bubbles between layers;
- spreading a thin layer of resin on the laminate before adding the foam or honeycomb, or better, a mat presoaked with resin to ensure good bonding.

Then begin again, systematically, on the other side of the sandwich core: mat, laminating and gel coat. Check that nothing has been forgotten between 9 and 16. The mould is now ready for curing. This requires vacuum control (less than 0.8 bar) and curing control (co-curing).

(b) Honeycombed sandwich, autoclave moulding by dry process

This is the high-performance type of sandwich structure with an excellent stiffness/strength to weight ratio, especially if carbon fibre epoxy resin prepreg skins are used with an aluminium honeycomb core.

Operating method
Pressure bag moulding is used (cf. Figure 4.9) with an autoclave. Note that low-flow adhesive films and prepregs are used to simplify the process by avoiding the need for bleeder cloth. The autoclave pressure P is smaller than 3 bars so as to avoid crushing the honeycomb.

Chronological control
Similar to vacuum bag moulding control (cf. Sections 4.1.3 and 4.2.2).

4.1.5 Filament winding

This process is an automatic winding process which results in a monolithic composite. The fibre orientation is excellent and the required strength and stiffness can be attained by choosing the appropriate fibre type and its placement. It is limited to cylindrical or symmetrical components.

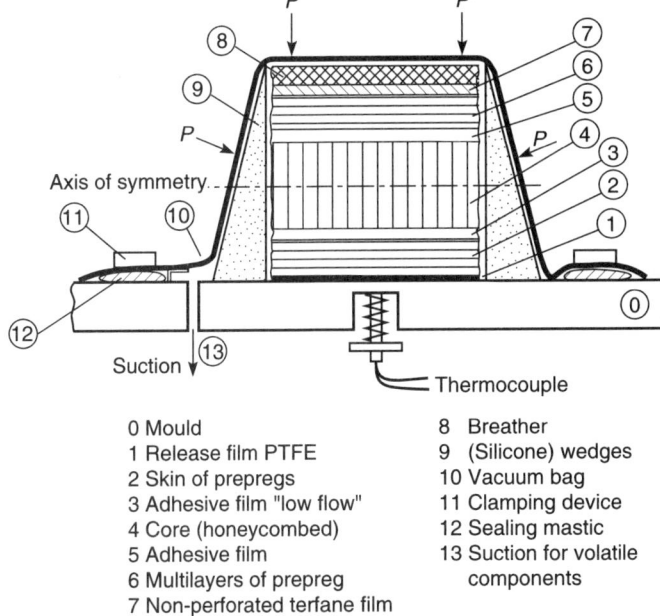

Figure 4.9 Pressure bag moulding in autoclave.

0 Mould	8 Breather
1 Release film PTFE	9 (Silicone) wedges
2 Skin of prepregs	10 Vacuum bag
3 Adhesive film "low flow"	11 Clamping device
4 Core (honeycombed)	12 Sealing mastic
5 Adhesive film	13 Suction for volatile
6 Multilayers of prepreg	components
7 Non-perforated terfane film	

(a) Principle

Filament winding (Figure 4.10) consists of regularly winding a reinforcing roving, impregnated with resin, around a usually cylindrical mandrel that can be removed, dismantled or destroyed when winding is complete and the resin cured. The impregnation must take place in a resin bath immediately before winding. Spools of presoaked roving may also be used.

Polymerization is achieved after winding, by placing the mandrel and winding in an oven or curing at ambient temperature. Alternatively a flexible bag may be placed over the winding and evacuated and the cure completed in an autoclave or oven.

(b) Control

The quality of winding depends on a number of factors which must be controlled in the course of the manufacturing process.

1. Surface quality of the mandrel. If this is made from several sections they must be assembled in such a way as to ensure a smooth surface.
2. Cleanness of the mandrel.
3. Spreading of the release agent.
4. If need be, the application of a gel coat depending on whether the exterior

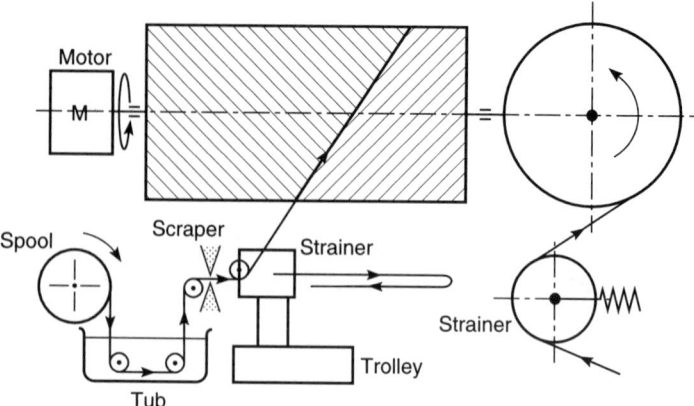

Figure 4.10 Filament winding.

or the interior of the component must have a good surface appearance. The nature, thixotropy and mass of the gel coat must be controlled and winding must be done on sticky gel coat.

5. Preparation of the resin: nature, amount of the hardener.
6. Viscosity of the resin; this is regulated by the temperature of the resin bath.
7. Resin impregnation ratio of the roving – this is regulated by a scraper blade as the roving comes out of the bath (or by the supplier in the case of preimpregnated rovings).
8. Winding pitch (trolley speed).
9. Winding speed.
10. Tension of rovings – this is regulated by a mechanical braking device on the feed spool.
11. Number of revolutions of the mandrel or number of wound layers.

Follows: control (cf. Section 4.2):

- of curing;
- of vacuum in the case of vacuum bag moulding;
- of autoclave pressure in the case of pressure bag moulding.

4.2 CONTROLLING THE CURING CYCLE

The quality of a composite depends on the following:

- the raw materials used: resins, fabrics, etc.;
- the curing cycle, i.e. the speed of temperature increase, the temperature of the

curing plateau, the time at which pressure is applied and the postcuring temperature and duration.

Good quality requires that:

- the curing cycle be perfected and optimized;
- this cycle be adhered to and every one of its stages checked during processing;
- the cured part be controlled at the end of processing.

4.2.1 Perfecting the curing cycle

In the case of heat polymerized composites, a curing simulator is required.

(a) Principle on which the curing simulator operates

Type one, the Vanhographe®, consists of two horizontal jaws which hold a flat specimen of laminate: (fabric + resin) or (mat + resin) or (SMC) or (prepreg). The whole is placed in a programmable thermal chamber. One of the jaws, actuated by a motor, will move in a sinusoidal manner, as shown in Figures 4.11 and 4.12. The receptor jaw is fastened to the base by spring plates and is influenced by the movement of the upper jaw and specimen. The amplitude of the movement of the receptor jaw is related to the stiffness of the specimen. This increases during curing. In the second generation simulator, special jaws cause the curing specimen to bend in alternate directions while the temperature is increased; this gives a measure of the glass transition temperature T_g.

Type two. A description of the apparatus can be found in the following sections:

- 2.2 Resin controls.
- 2.3 Prepreg control.

The specimen is placed between vertical jaws and twisted back and forth.

(b) Determining the speed of heating

Several identical samples must be tested and the change of stiffness, R, in the course of the curing process, determined for temperature rises of 1, 2, 3, 4, 5 °C per minute, as shown in Figure 4.13.

A high heating rate rapidly reduces resin viscosity, and the fabric should be well impregnated. However, resin leakage may occur and the part being moulded may eventually be starved of resin or the composite may be inhomogeneous.

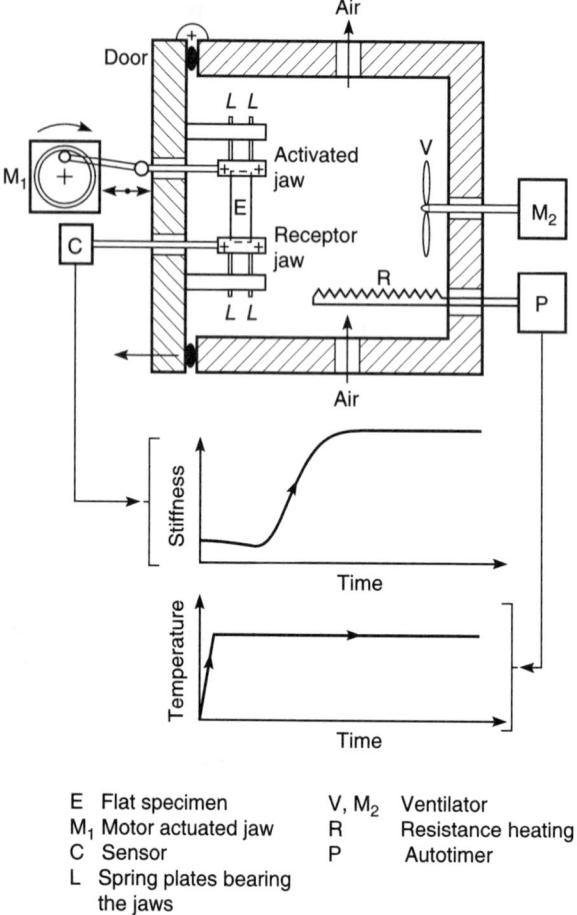

Figure 4.11 Schematic view of Vanhographe®.

E	Flat specimen	V, M₂	Ventilator
M₁	Motor actuated jaw	R	Resistance heating
C	Sensor	P	Autotimer
L	Spring plates bearing the jaws		

A low heating rate gives a higher resin viscosity. As a consequence, impregnation may be reduced especially for interior fabrics.

Gelling is a slow and progressive process. To ensure a good flow of resin in these circumstances, it would then be necessary to increase moulding pressure.

(c) Determining the temperature of the curing plateau

Once the heating rate has been selected (several °C per min) (see Section 4.2.1), the test is carried out on identical samples 1, 2, 3,..., n, which are cured at

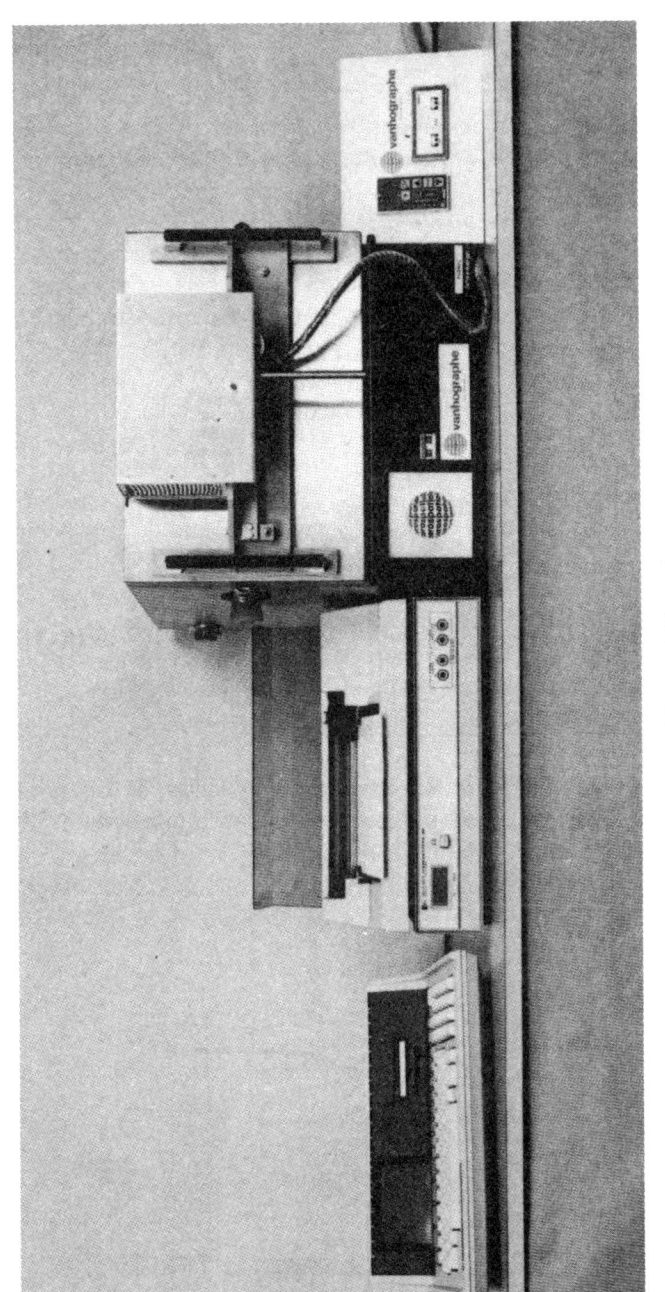

Figure 4.12

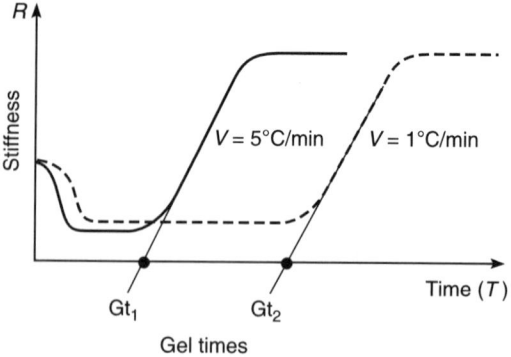

Figure 4.13

increasing temperatures T_1, T_2, T_3,..., T_n. The duration t of the curing process should be adequate (four hours, for instance) and be fixed beforehand. After curing, the samples are tested to determine the glass transition temperature T_g. The values obtained as a function of the curing temperature T_{g1}, T_{g2}, T_{g3},..., T_{gn} are noted, as in Figure 4.14.

The curing temperature which produces the highest glass transition T_g without, however, reaching the temperature at which the resin starts deteriorating, or causing the surface of the composite material to be damaged, should be adopted.

A high T_g shows that a high degree of cross-linking has occurred. The composite produced will have excellent dimensional stability though it may be more brittle than if cured at a lower temperature.

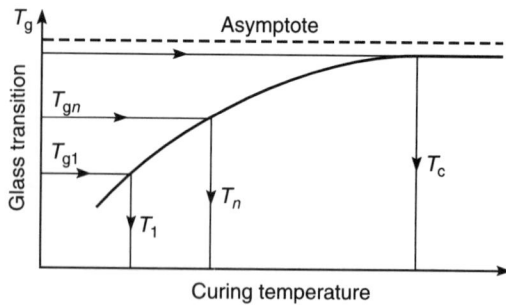

Figure 4.14

(d) Selecting the ideal time for applying pressure to the mould or creating a vacuum in the bag

Selection will be based on (see Section 2.2, Resin controls).

A specimen is cured according to the temperature cycle chosen previously (speed V, isothermal temperature T). Its stiffness is determined relative to time, as in Figure 4.15.

Note: Do not confuse the T_g glass transition temperature with Gt (gel time).

In the case of a reactive resin, the (R, t) stiffness–time curve is very steep. The interval Δt between the beginning and the end of the gelling process is short. Pressure P must be applied shortly before the beginning of gelling (Gt) so that the resin is able to flow between the layers of the fabric, as in Figure 4.16.

Once the gelling process has begun, the viscosity of the resin increases rapidly. Consequently, the flow of resin between the layers may be inadequate for proper wet-out resulting in a poor laminate. In the case of a resin with low reactivity, gelling is very slow. If pressure were applied before Gt, the resin would leak out of the mould or from the laminate. Pressure must be applied when the gel time has been reached or shortly afterwards.

(e) Determining demoulding time

The ideal, economic, demoulding time corresponds to the time when the stiffness of the composite has ceased to increase. When this is so the stiffness–time curve is almost horizontal. It is easier to detect this on a computer-derived curve of dR/dt as the latter is zero or close to zero by then (see Figure 4.17).

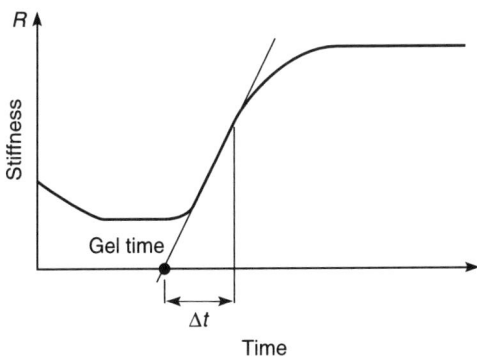

Figure 4.15

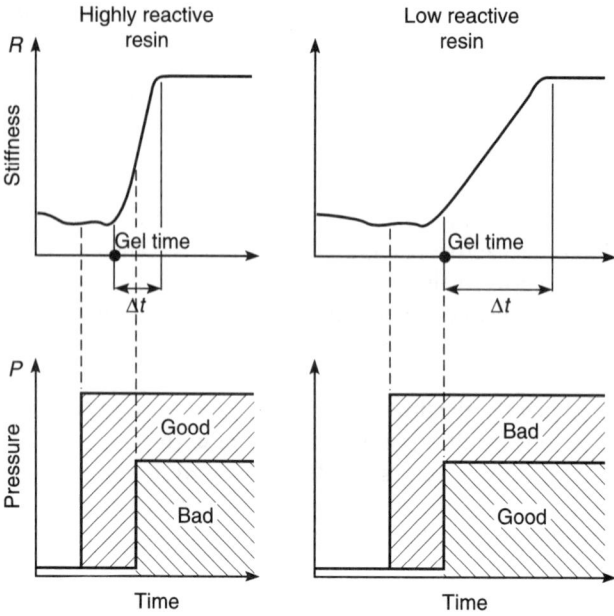

Figure 4.16

It must be noted that at the demoulding time, t_D, the part is not completely cured. This can be shown using the Kinemat®. The stressing cycle is stopped halfway through to allow the material to relax. It can be seen from Figure 4.18 that the curve for the relaxed torque (C_r) differs from that of the instantaneously transmitted torque (C_0) showing that cross-linking is not complete.

The polymerization process is complete at time t_p when the relaxation curve (C_r) becomes horizontal or, when the derived slope dC_r/dt is zero.

At demoulding time, (t_D), the material is viscoelastic. The loss of shape due to the stress applied to the part during demoulding will be absorbed by viscoelasticity. The part will retrieve its original shape once it has been taken out of the mould and the demoulding process should cause no damage.

(f) Post-curing time and temperature

The part may be demoulded at time t_p which corresponds to the completion of polymerization. No post-cure is necessary. However the mould is being used for longer than necessary and it is not possible to use it for another component. Thus there is a loss of productivity.

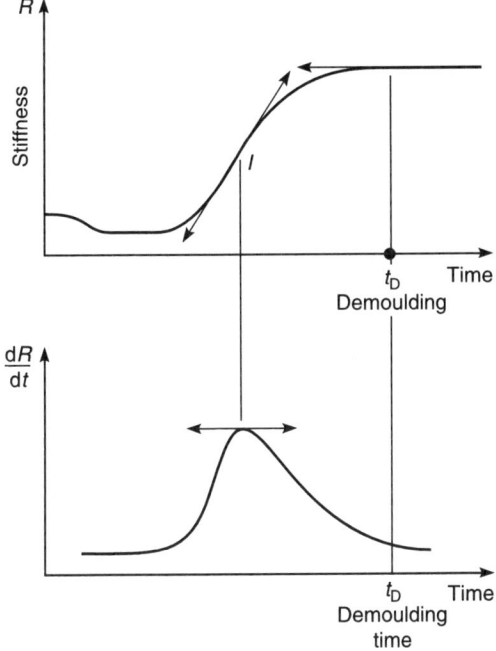

Figure 4.17

If the part is demoulded at time t_D (see Section 4.2.1(e)), it is not fully cured and must therefore be post-cured. The minimum post curing time will be $\Delta t = t_p - t_D$, an interval which can be determined from the stiffness curves (see Section 4.2.1(e)). The temperature of post-curing (T_{post}) is the same as that of curing (T_c).

To speed up the post-curing process, or to ensure more thorough cross-linking, a post-curing temperature higher than that of curing may be used. In this particular case, care must be taken to use a temperature (T_{post}) lower than T_g, the glass transition temperature of the resin cured at T_c, that is,

$$T_c \leqslant T_{post} < T_g < T_{\text{resin degradation}}.$$

If $T_{post} \geqslant T_g$ were adopted, large and heavy composite parts, which were not post-cured in a jig which controlled shrinkage, might deform under their own weight.

(g) Perfecting the curing cycle using dynamic viscoelastic analysis

The rheological behaviour of resins varies according to the temperatures to which they are subjected. Generally they are visco-elastic rather than elastic.

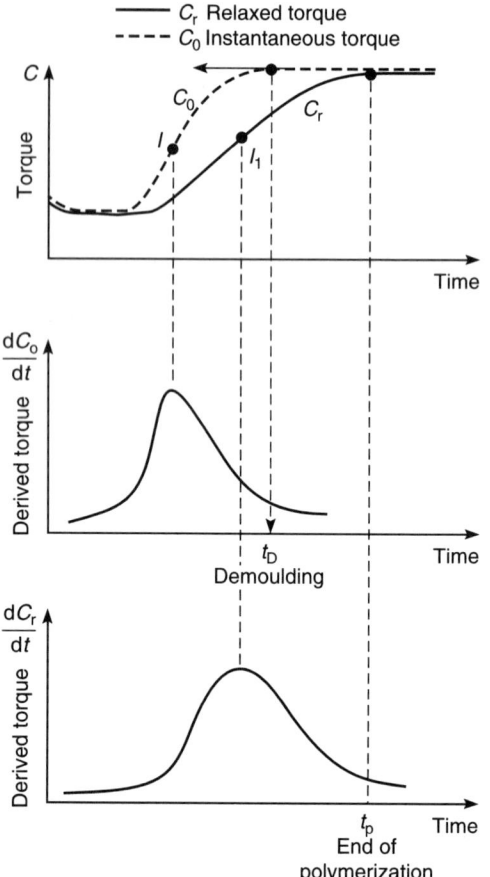

Figure 4.18

Whenever a non-elastic material is distorted, part of the energy generated in the deformation process is released in the form of heat, and, if the material is subjected to an oscillating load, the remainder in the form of vibrations whose frequency v is proportional to the stiffness of the material. The stiffness of a resin, heated slowly at $2\,°C$ per minute, is related to the temperature. The change of stiffness E follows the same pattern as that of viscosity η or frequency v, as shown in Figure 4.19.

The curve obtained is the same as that given by a curing simulator. It should be noted that the change in E is rapid, and that the interval between the time T_1, when it is a minimum, and Gt, the gel time, is very short indeed, approximately a quarter of an hour for an epoxy resin.

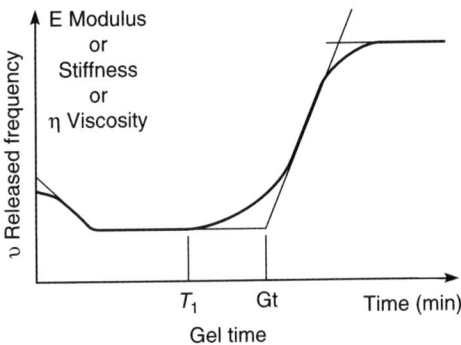

Figure 4.19

It is helpful to cure at a lower temperature than that of polymerization in order to regulate the viscosity of the resin before applying pressure.

During the viscous stage (to Gt), the resin flows readily between the fibres under little pressure. There is a danger that the viscosity may be so low that resin will drain from the component.

During the latter stage ($t >$ Gt), the resin will not flow within the mould unless very high pressures are applied. This is virtually impossible in most industrial plants. For instance, the maximum pressure that can be obtained in an autoclave is of the order of 10 bars. With a very viscous resin under these circumstances, not all the fibres will be properly impregnated. The part produced will be defective, and delamination might occur later associated with poorly impregnated areas or pores in the matrix.

(h) Recommended curing cycles

Depending on the change of resin stiffness during the time selected for a given curing temperature, the ideal cycle is as follows.

1. Before reaching the curing (polymerization) temperature, the laminate should be preheated for a certain time. Polymerization will be initiated during this preheating stage, which will also be used to regulate viscosity of resin before pressure is applied. Moreover, this preheating stage will produce a uniform temperature within the part, which is very desirable particularly in the case of large, thick parts moulded in equipment with a high thermal inertia or heat/temperature lag.
2. Pressure should be applied shortly before the gel point.

Autoclave moulding

The curing cycle of an epoxy resin (Ciba-Geigy 914) in an autoclave will be given as an example. First, the carbon fibre resin prepreg laminate is placed in a vacuum bag to assist in the elimination of volatile products, then it is placed in the autoclave.

Note that if a 'low-flow' resin, which has a higher viscosity than a 'controlled flow' resin, is used, there will be no risk of the resin flowing and escaping from the mould at the beginning of the process. It will therefore be possible to apply pressure earlier, as seen in Figure 4.20.

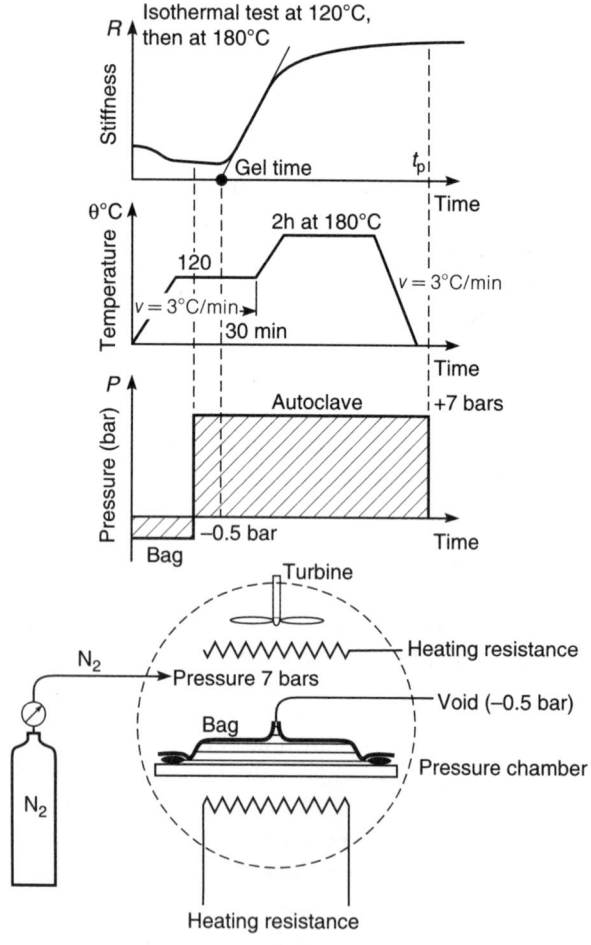

Figure 4.20 Autoclave moulding.

Vacuum bag moulding
In the case of thick parts where the heat released by the reaction is not very high and cannot therefore degrade, epoxy resin is considered, as on Figure 4.21. In the case of pressure bag moulding, either pressure or a vacuum is applied shortly before the gelling of the resin begins. This is a general rule whatever the thermal cycle recommended by the manufacturer may be.

Method for thick components
In order to avoid an excessive exotherm which could degrade the resin, it is strongly recommended that the temperature is maintained at 80 °C for two to four hours before increasing it to 120–160 °C, as shown in Figure 4.22.

(h) Methods for composites cured at room temperature without the application of pressure

Contact moulding
The curing simulator can be used at room temperature to determine t_D, the demoulding time, and t_p, polymerization or curing time, under these conditions.

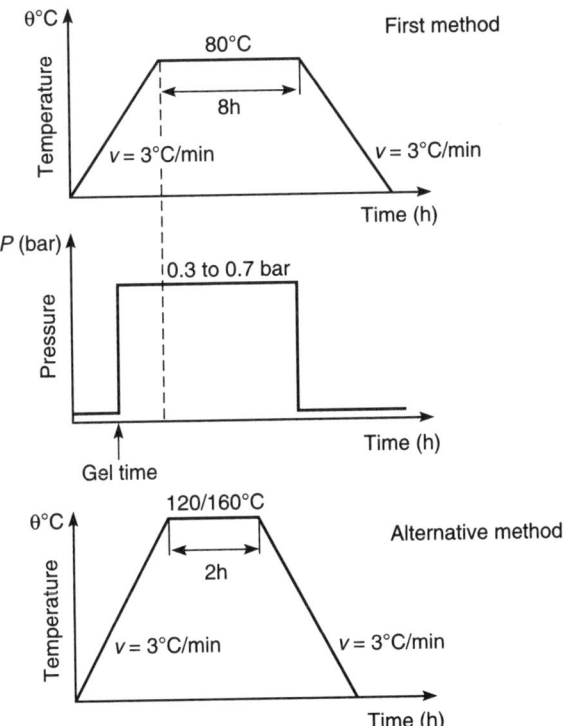

Figure 4.21 Vacuum bag moulding.

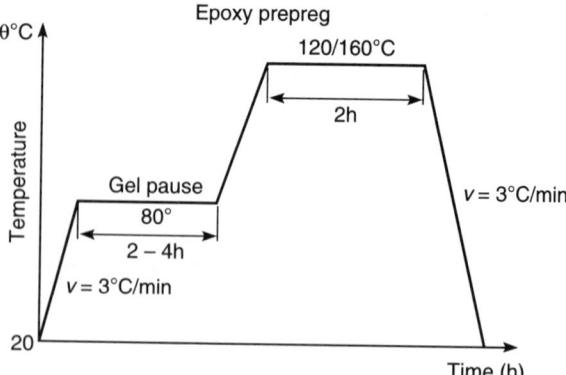

Figure 4.22 Vacuum bag moulding of thick part.

Examples include: Rezolin 661 epoxy resin (France) cured at room temperature $t_D = 12$–16 hours at 25 °C, $t_p = 5$ days at 25 °C; and Ciba-Geigy XB 3052 epoxy resin (Switzerland) suitable for cold and hot curing, $t_D = 16$ hours at 25 °C, $t_p = 7$ days at 25 °C.

4.2.2 Operating control of the cycle

The cycle that has been perfected by the laboratory, or possibly the one recommended by the resin or the prepreg manufacturer, must be observed during processing in the workshop. The workshop personnel are in charge of the following:

- mould heating;
- ovens;
- autoclaves;
- vacuum pumps;
- presses.

They will programme the temperature and pressure cycles and ensure that the cure schedule is followed in detail. This is essential for the component to be successfully processed.

Increasingly, curing cycles are being automatically controlled through microprocessors. This allows the following three different functions to be covered:

- running of curing;
- supervision and alarm in case of any deviation from the selected programme;
- data collection, storage and processing.

The automatic equipment that runs the cycle may consist of the following:

- input – cassette player for information/instruction storage, thermocouple sensors, pressure sensors;
- output – power relays, electromagnetic valves, control motors, alarm indicators.

The functions of the various automatic controllers are the following:

- to read the pressure/temperature schedule;
- to compare the required procedure with that indicated by the sensors;
- to regulate the process by alleviating both pressure and temperature in real time.

4.2.3 Post-curing control

It is difficult to achieve complete polymerization. In excess of 95% cross-linking is indicative of satisfactory curing and post-curing processes.

(a) Defining the glass transition temperature

To determine whether an item has been properly cured, the glass transition temperature (T_g) of a sample is determined with equipment designed to study the rheological behaviour of the laminate or the resin.

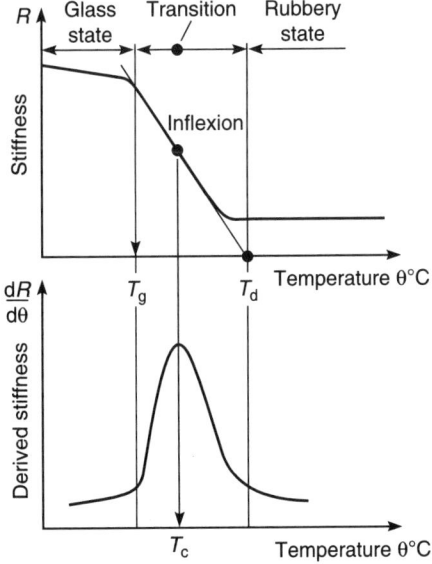

Figure 4.23 Curing simulator analysis.

Analysis of laminate sample with the curing simulator

A laminate sample is slowly heated ($v = 2\,°C/min$) in the curing simulator. The change in its stiffness, R, with temperature, θ, or of the derivative $dR/d\theta$ is monitored, as in Figure 4.23.

Glass transition is an interval which:

- begins at T_g;
- reaches a maximum at T_c;
- ends at T_d.

The value of the glass transition temperature indicates the degree of cross-linkage within the material. T_g increases with the cure temperature and duration of curing. A 100% polymerized composite would have a glass transition identical to the decomposition temperature of the resin.

The loss of stiffness ΔR_1 or ΔR_2 on heating the sample is greater the lower the degree of cross-linking in the resin matrix, as in Figure 4.24.

When used at a temperature $\theta > T_g$ the matrix becomes rubbery and the item may lose its shape because of residual moulding stresses or applied stress.

Analysis of a laminate or resin sample with a torsional pendulum or DMA

The variation of shear modulus (G) with increasing temperature can be measured with a torsional pendulum. Alternatively a DMA (Figure 4.25) may be used to study the relaxation modulus (E') as the temperature increases.

During the glass transition the modulus undergoes a sharp fall, as shown in Figure 4.26. Results obtained for epoxy resin are shown in Figure 4.27.

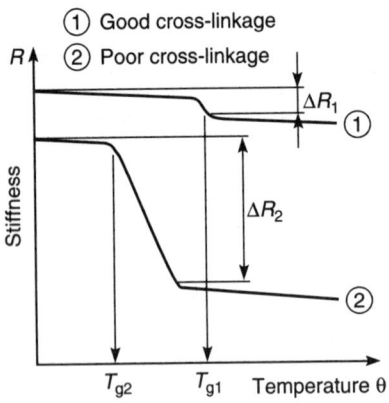

Figure 4.24

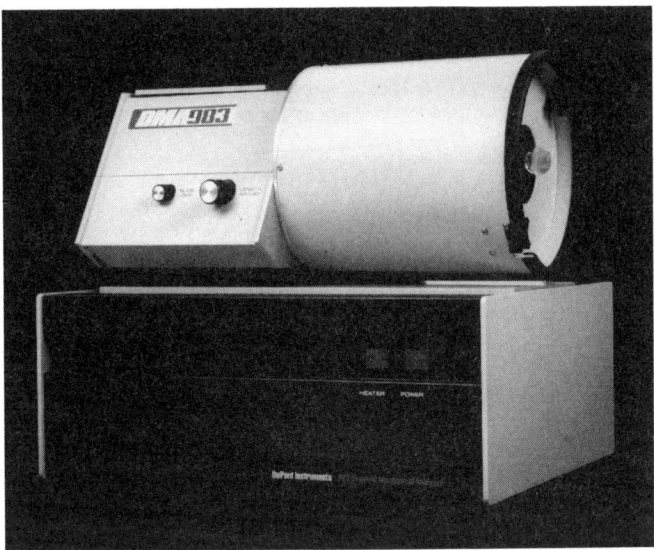

Figure 4.25 Viscoelastic analyser (differential mechanical analyser) (du Pont de Nemours).

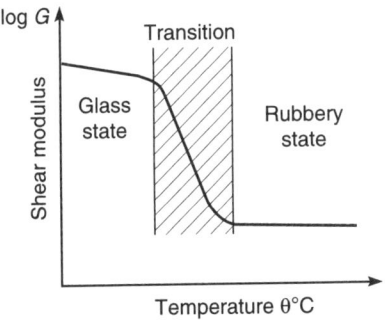

Figure 4.26

Analysis of a laminate sample with a thermodilatometer
In the vicinity of the glass transition, the coefficient of thermal expansion of the resin, or its specific volume, changes markedly. However methods based on this are less sensitive than DSC or DMA in detecting the glass transition temperature of a cross-linked laminate. Further cross-linking due to extra heating can also be detected, as shown in Figure 4.28.

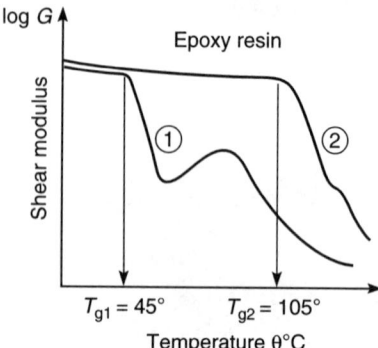

Figure 4.27

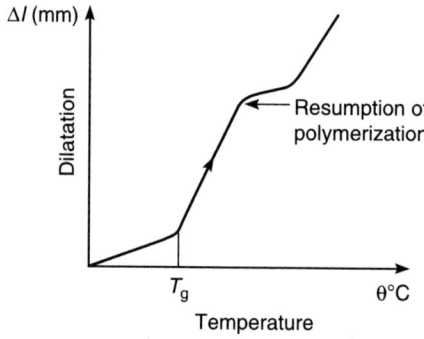

Figure 4.28 Controlling curing with a dilatometer.

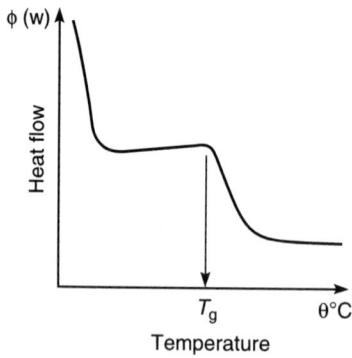

Figure 4.29 DSC analysis.

DSC analysis
A small piece of polymerized laminate is put in the DSC and the temperature is raised rapidly (10 °C/min). The heat absorbed by the sample is recorded as a function of temperature and undergoes a sharp change at the glass transition temperature (T_g), as in Figure 4.29.

(b) Checking for incomplete curing – tests carried out on moulded items

Samples taken from items moulded in the workshop are used.

Rapid method using a curing simulator
A small sample taken from the moulded item or an extra piece produced at the same time is placed in the curing simulator. The sample is submitted to a rapid temperature increase (2–3 °C/min) until it reaches $\theta > T_g$ as shown in Figure 4.30 (T_g having been determined in a preliminary test). θ must be lower than the decomposition temperature of the resin.

First, the stiffness (R) of the item is reduced because the resin becomes rubbery at temperature $\theta > T_g$. Heating above T_g causes further polymerization, and increased stiffness. Once the specimen is thoroughly cross-linked, its stiffness is higher than at the beginning of the process.

The method described above is a quick one which can be completed in less than an hour.

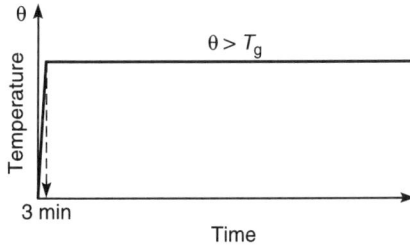

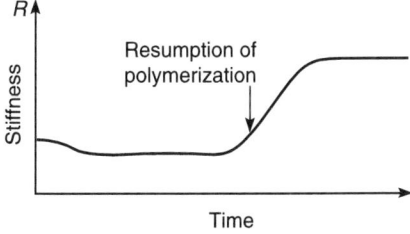

Figure 4.30 Curing test in simulator: fast method.

Slow method using a curing simulator

The purpose of this is to determine whether curing or post-curing times have been long enough. A sample taken from a cured item is placed in the curing simulator at a temperature of θ °C, the temperature is then increased slowly to that of curing or post-curing. The change of stiffness is recorded over a period of several hours, as shown in Figure 4.31.

If stiffness increases during the test, it is a sign that the composite was poorly cured. The cross-linking of the item can be improved by putting it in an oven for several hours at the appropriate temperature.

Differential scanning calorimetry method (DSC)

This is carried out using a differential scanning calorimeter (DSC), as illustrated in Figures 4.32 and 4.33.

A small sample of material (mass M (g)) is placed in a DSC, and slowly heated up. If the material is insufficiently cured, additional polymerization will take place producing a characteristic thermal flux. If the temperature is raised sufficiently, the resin will begin to decompose, as shown in Figure 4.34.

The area S under the polymerization peak is proportional to the heat produced by the additional cross-linking. This spectrum can be compared with that furnished by the same mass, M, of uncured material (prepreg or resin preimpregnated fabric) which will yield a peak of area $S_0 \gg S$. The S/S_0 ratio indicates the degree to which the poorly cured laminate has cross-linked, as shown in Figure 4.35.

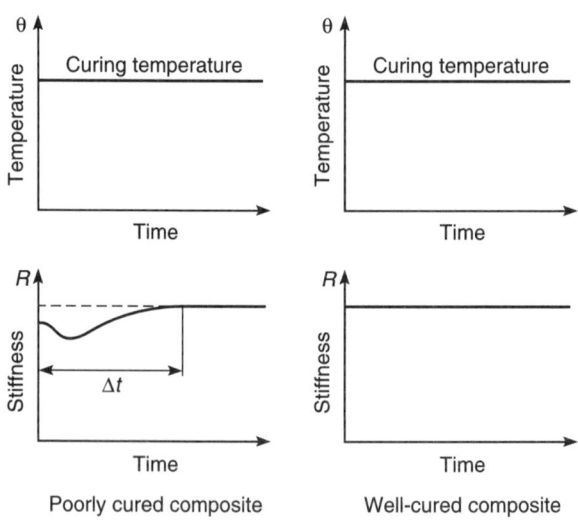

Figure 4.31 Curing test in simulator: slow method.

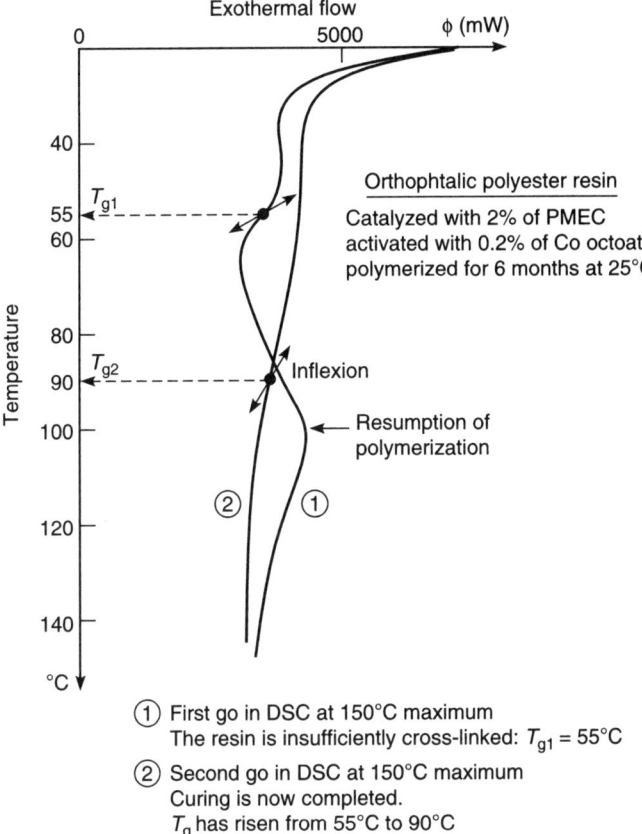

Figure 4.32 DSC curing test.

Thermodilatometry method
It is possible to measure the dilatation or volume or length change of a small
bar of laminate as the temperature varies. The results will be characteristic
of the following (see Figure 4.36):

- glass transition temperature, T_g;
- possible resumption of polymerization;
- resin degradation at T_D.

The glass transition temperature is, however, harder to determine for a
composite sample than in a bar of pure resin.
In the case of a UD laminate, results will vary depending on the fibre
direction. For maximum sensitivity the fibres should be transverse to the

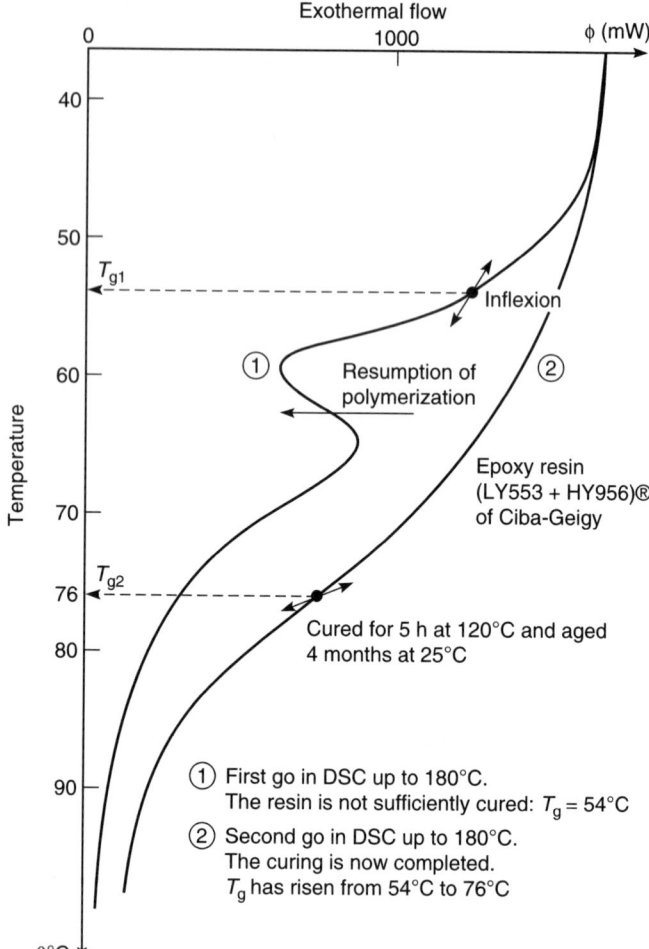

Figure 4.33 DSC curing test.

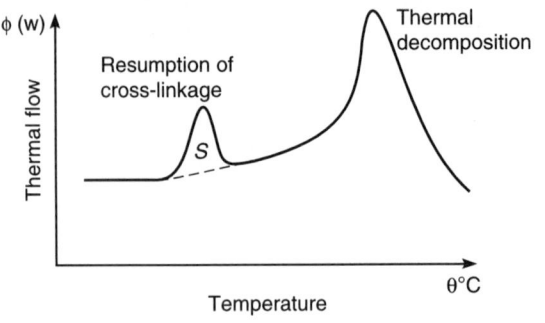

Figure 4.34 DSC curing test.

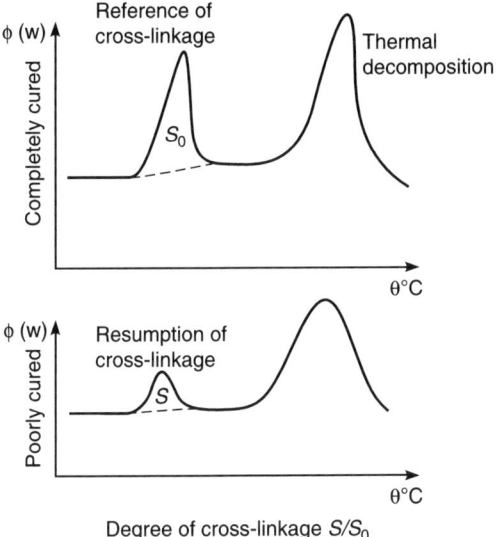

Figure 4.35

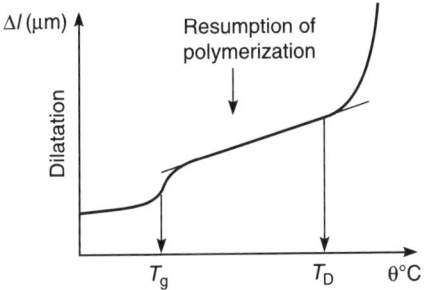

Figure 4.36 Thermodilatometry curing test.

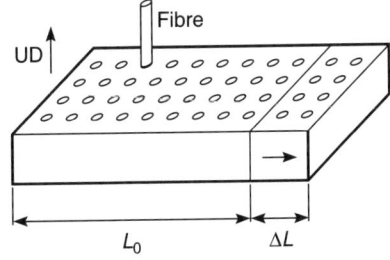

Figure 4.37

long axis of the specimen, as shown in Figure 4.37. Figure 4.38 shows the results obtained for well and poorly cured laminates.

If the test is carried out on a pseudo-isotropic composite, the results will vary with the lay-up. It is, however, easy to tell insufficiently cured laminates from others, as in Figure 4.39.

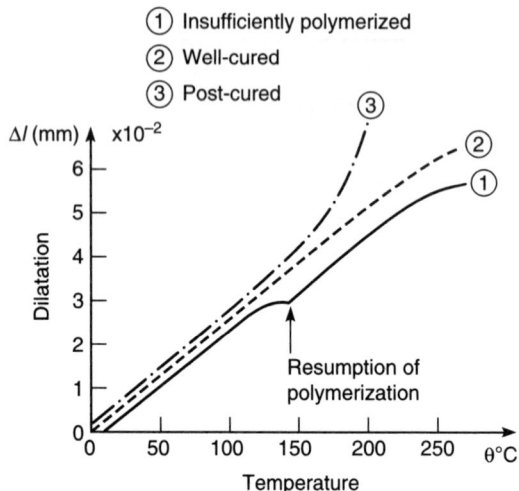

Figure 4.38 Thermodilatometric analysis of carbon-epoxy UD laminate (source: Aérospatiale, France).

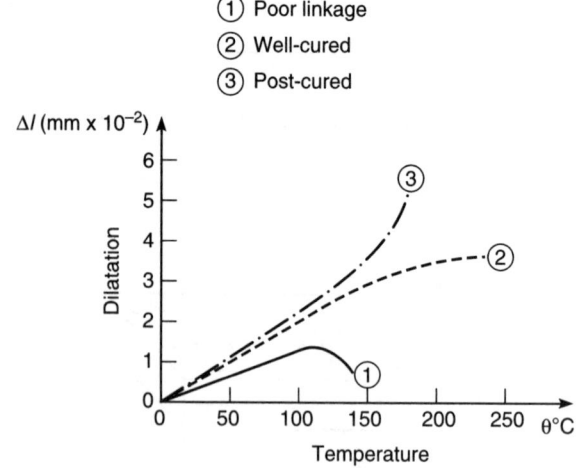

Figure 4.39 Thermodilatometric analysis on quasi-isotropic carbon-epoxy.

4.3 CONTROL OF THE CURING CYCLE OF SANDWICH SKIN–CORE ADHESION

4.3.1 The problem

The problem is to determine the curing of a film of adhesion placed between the skin and core of a sandwich structure, as shown in Figure 4.40. Bonding can be achieved in two different ways: either the adhesive is spread between an already polymerized skin and a core of some kind (balsa, honeycomb, foam); or, the skin and adhesive film are stacked on the core of the sandwich and cured together (see Figure 4.41).

Sometimes the surface of the foam core is first coated with a mortar (Figure 4.42) made up of laminating resin and glass microballs to smooth it over by filling in the pores and the skin is applied to this and cured in the normal manner.

In all cases the curing parameters (that is, curing temperatures and applied pressure) must be checked. It is essential to monitor the temperature reached at the joint, not the temperature within the oven or autoclave.

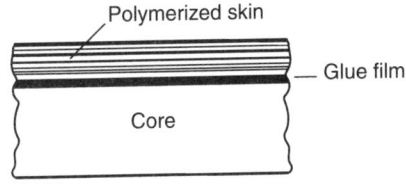

Figure 4.40

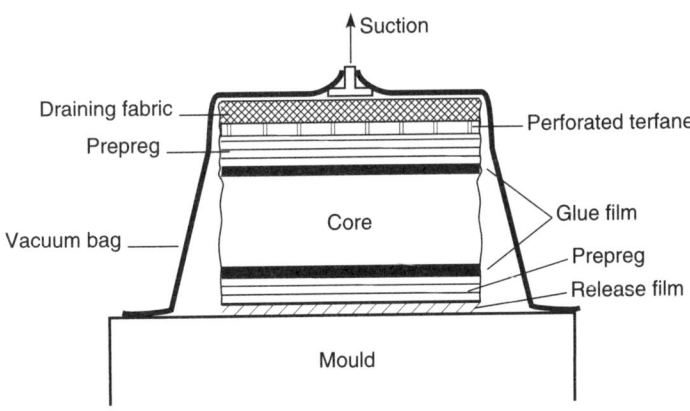

Figure 4.41

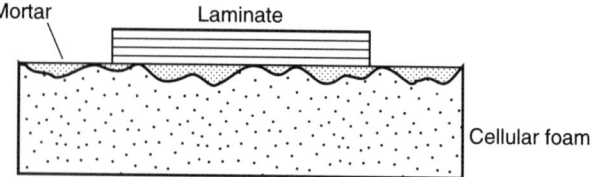

Figure 4.42

4.3.2 Polymerization of the adhesive film

X-rays or radiation might be used, but heat is by far the most common means. Heat is produced:

- by convection (in an oven or an autoclave);
- by conduction (heating presses, heating moulds or a heating mattress);
- by IR – this allows localized bonding;
- by HF current – this requires thick, insulating, supports;
- by microwaves.

4.3.3 How to apply pressure

This can be done by:

- applying a weight to the mould;
- using spring loading on the two parts of the mould to give a tight seal;
- hydraulic press;
- jacks;
- vacuum bag;
- autoclave and pneumatic bag.

4.3.4 Curing cycle of the adhesive

As is the case with any thermosetting resin, the curing cycle for the adhesive is developed by the supplier (see Section 4.2). Means of doing this include:

- DSC thermal analysis;
- DMA dynamic mechanical analysis;
- curing simulator (Kinemat® or Vanhographe®).

When simulating curing, the nature of the curing equipment used in the workshop must be taken into account:

- heating power of the oven, press, autoclave and other pieces of equipment;
- thermal inertia of the equipment used.

This allows the rate of heating in the simulation to be set at a realistic value.

Tests carried out in a curing simulator show how the stiffness (or the viscosity) of the adhesive film varies with time and temperature and give the gel time, Gt, as shown in Figure 4.43. The pressure must be applied shortly before Gt. Adhesives are 'low-flow' resins.

An example of the polymerization or curing cycle of an adhesive in an autoclave is shown in Figure 4.44.

Whenever adhesives are used which may release volatile matter, thus creating bubbles which will remain trapped within the adhesive film, the whole assembly should be placed inside a vacuum bag under a negative pressure P_2 of the order of 0.1 bar. This will remove the volatile products and avoid the formation of bubbles and voids. Moreover, the decreased pressure will increase the wetting power of the adhesive and allow it to fill in the cells of the foam, balsa or honeycomb more efficiently.

Autoclaves may be pressurized with compressed air or, better, nitrogen. When bonding, the maximum pressure that can be applied depends on the crushing resistance of the sandwich core (see Section 2.5), and on the flow of the glue film. Usually the pressure should not exceed $P_1 = 1$–3 bars, and should be applied shortly before Gt.

The heating/cooling speed must be low. The larger the part to be bonded, the lower the speed used. This is very important if the viscosity of the adhesive is to be controlled. The cooling rate must be low to ensure relaxation of residual stresses after the polymerization. Moreover, a slow cooling rate allows for differential expansion between the mould and the cured component.

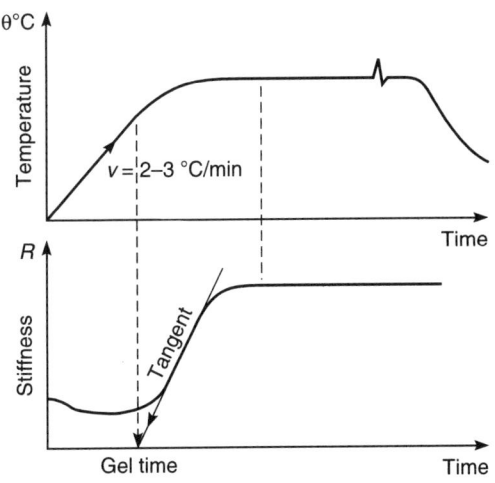

Figure 4.43

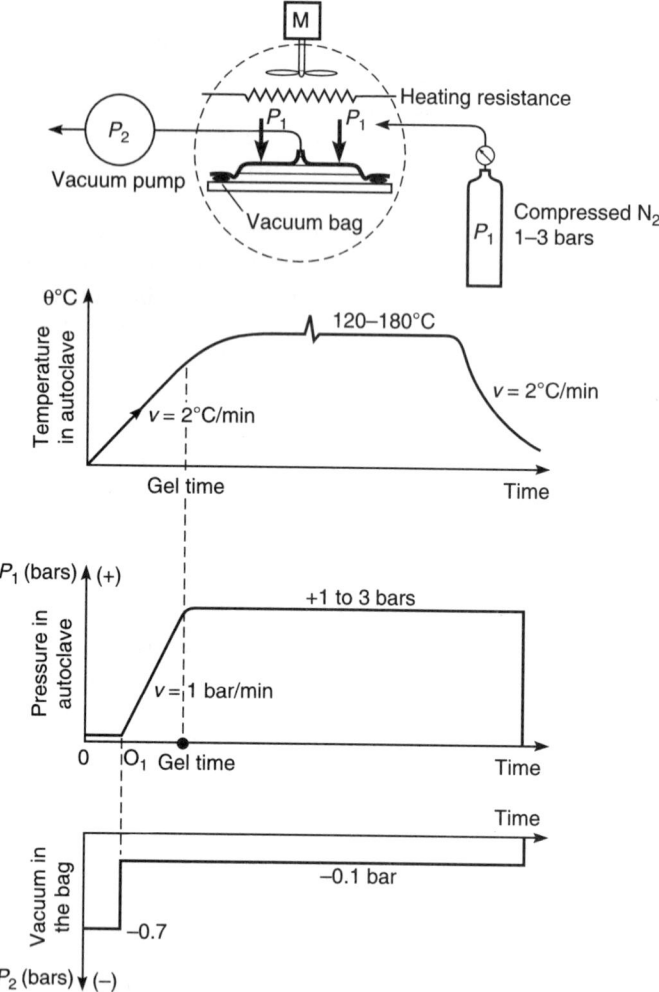

Figure 4.44 Curing cycle in an autoclave.

The moulding equipment is generally made of steel or aluminium alloy, 2017 or 2024, and its coefficient of expansion is very different from that of the composite (resin, adhesive, fibre):

Steel $\alpha = 12 \times 10^{-6}/°C$
Aluminium $\alpha = 24 \times 10^{-6}/°C$
Resins, adhesives $\alpha = 40$ to $80 \times 10^{-6}/°C$
Fibres: carbon $-0.5 \times 10^{-6}/°C$.
         glass $+5 \times 10^{-6}/°C$

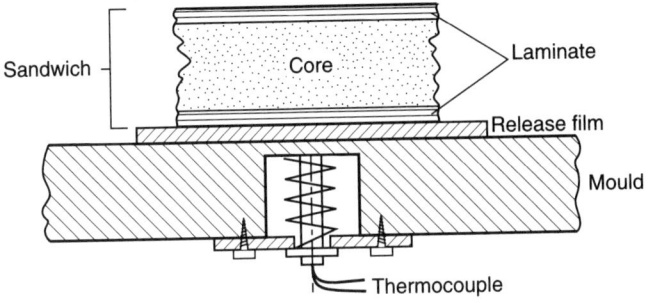

Figure 4.45

In the course of the cooling process the shrinkage of the core and skin of the sandwich structure of the adhesive and of the mould are all different. This leads to internal stresses which are partially reduced through relaxation in the structure.

Adhesion of the skin to the mould hampers the shrinkage process, and can be avoided by placing a film of PTFE (Teflon®) between the sandwich skin and the mould.

Note that the heating of the assembly to be bonded must, if at all possible, be uniform. If heating only occurs on one side, the moulded part would be distorted.

4.3.5 Cycle control

The cycle, having been established, must be monitored. For this purpose, the fabrication equipment should be fitted with:

- pressure sensors: manometers;
- temperature sensors: thermocouples;
- recorders to note the pressure and temperature cycles.

Monitoring the curing cycle is extremely important. The cycle recommended by the manufacturer or laboratory should be followed as closely as possible and any fault or deviation reported.

The thermocouples should be fastened to the mould and be located as close as possible to the composite, as shown in Figure 4.45. Consequently the temperature recorded is slightly higher than in the adhesive film.

The trend now is towards monitoring cycles using sensors and microprocessors. This allows both the automatic adjustment of the curing cycle and can activate an alarm in the event of any fault.

The storing of data/readings makes it possible subsequently to verify that the component has been properly cured and is of great help in proper quality control.

5.1 DESTRUCTIVE TESTING OF SAMPLES OR TEST PIECES

5.1.1 List of tests and related standards

Once the product has been finished, the laminate can be tested as follows. Use sample test piece(s) cut from one of a series of articles; the selected article must subsequently be scrapped, however. This technique can be used in the case of series of cheap and small articles.

Use extra test piece(s). This technique is used in the case of large, expensive artefacts or one-off fabrications. The test piece(s) are cut from an over-size artefact.

Another method is to move the moulding and a small test plaque at the same time and by the same method (i.e. exactly the same temperature, time, pressure cycle).

The tests performed on the samples are essentially laboratory measurements. They cover the mechanical, physical and chemical properties of the material.

(a) Physical tests

- Determination of volume and weight fraction of the fibre reinforcement.
- Determination of the volume and weight ratio of the plastic matrix.
- Determination of the specific gravity or density of the complete laminate.
- Determination of the void content of the laminate.
- Evaluating the state of cure of the laminate.

(b) Mechanical tests

There are many possible mechanical tests and they must be chosen carefully to limit costs. In order that the results have some statistical significance, at least five test determinations are required, per property, per material. Usually the static flexural or tensile strength is sufficient. The following list is more exhaustive.

- Determination of tensile characteristics (strength, stiffness and strain to failure).
- Determination of flexural characteristics (strength, stiffness and strain to failure).
- Determination of interlaminar shear stress.
- Determination of impact strength.
- Determination of compression strength, modulus and strain to failure.
- Measurement of hardness.

(c) Related standards

NFT 57-102, NFT 57-571, NFT 57-518, NFT 57-557, ISO 1172, EN 60, DIN 53395. Determination of the glass fibre content by ignition.

NFT 57-608. Textile glass fibre prepregs – fibre and resin content – by chemical digestion.

EN 2564, ASTM D3171-73. Determination of carbon fibre content.

NFT 57-101, NFT 57-301, L17-410, project EN 2561, ASTM D3039, DIN 53 455, BS 2782 part 10 method 1003. Determination of tensile properties.

NFT 57-151, NFT 57-152, NFT 57-153, NFT 57-300. Preparation of flat sheets or laminated plates or panels for test specimens (tensile and other purposes).

NFT 57-105, NFT 57-302, L17-411, EN 63, project EN 2562, ISO 178. Flexural test.

ISO 4585, NFT 57-104, NFT 57-303, L17-701, L17412, EN 2377, project EN 2563, ASTM 2344. Apparent interlaminar shear strength.

NFT 57-108, ISO 179. Charpy impact strength test.

NFT 57-103, ISO 8515, DIN 53-454. Compression test parallel to the plane of lamination.

NFT 57-106, EN 59. Barcol hardness.

5.1.2 Determination of fibre content

(a) Importance of the test

The mechanical characteristics (tensile strength and modulus) of a laminate for any type of lay-up (unidirectional or pseudo-isotropic) are directly related to the reinforcing fibre content. It is therefore important that the fibre ratio should be monitored regularly during the manufacturing process and so kept at a constant value during mass production. See Figure 5.1, where V_F % is the volume percentage of reinforcing fibre, F refers to the fibre, M refers to the matrix.

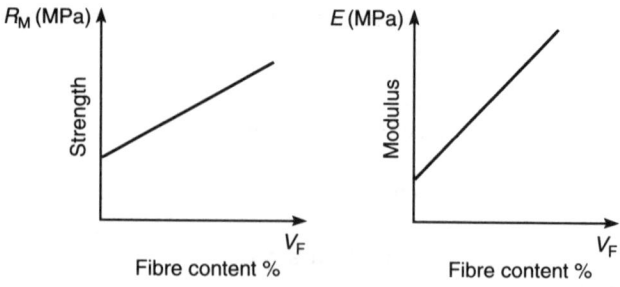

Figure 5.1

(b) Significant relations

We denote percentage weight reinforcement content as:

$$M_F = \frac{\text{mass of the reinforcement contained in the laminate}}{\text{total mass of the laminate}} \times 100.$$

The percentage weight resin content is:

$$M_m = \frac{\text{mass of resin in the laminate}}{\text{total mass of the laminate}} \times 100.$$

Further, if no voids are present $M_F (\%) + M_m (\%) = 100\%$ or $M_m = 1 - M_F$.
In the same way we denote the percentage volume reinforcement content as:

$$V_F = \frac{\text{volume of reinforcement}}{\text{total volume}} \times 100$$

and $V_m = 1 - V_F$.
Weight and volume fractions or percentages are related as follows:

$$V_F = \frac{m_F/\rho_F}{m_F/\rho_F + M_m/\rho_m} \quad \text{and} \quad M_F = \frac{V_F \times \rho_F}{(V_F \times \rho_F) + (V_m \times \rho_m)},$$

ρ_F and ρ_m being the specific gravity of the fibres and matrix. The specific gravity ρ_s of the laminate can be related to the fibres and matrix content by the following relation:

$$\rho_s = (\rho_F \times V_F) + (\rho_m \times V_m).$$

(c) Determination of glass fibre content

Standards: NFT 57-102, ISO 1172, EN 60, DIN 53 395, BS 2782 part 10 method 1002.

In a glass fibre resin laminate, the resin is burnt off leaving the fibre behind. Refer to the following standards: NFT 57-518, ISO 1172, NFT 57-571, NFT 57-608.

A piece of composite conditioned at 23 °C and 50% RH is weighed accurately, total mass $= m_t$. It is placed in a heat-resistant crucible made of porcelain, the weight of which, m_0, has been determined. The crucible and the composite are placed in an oven at 625 °C. The laminate may be set on fire with a torch. When combustion is complete (there is no longer any flame), glass fabric and charred residue remains.

The crucible and contents are left in the oven at 625 °C until the charred residue has completely disappeared (up to 12 hours at 625 °C), leaving only glass fibre. The crucible and its contents are cooled in a desiccator to stop the fibres absorbing moisture and reweighed, m_1, when cool.

m_t = mass of the glass + mass of the resin = mass of the laminate.
m_0 = mass of the crucible.
m_1 = mass of the crucible + mass of residual glass.

Hence, the mass of the glass is $m_g = m_1 - m_0$ and the glass content of the laminate as a percentage of the total mass is:

$$M_F = \frac{m_1 - m_0}{m_t} \text{ (in %)}.$$

Note that to achieve greater accuracy, four or five samples can be conditioned and tested together. Combustion at 625 °C removes any size on the glass but the mass of this is negligible anyway. If the resin contains non-combustible and non-volatile fillers, for instance thixotropic silica SiO_2, these are included in the mass $m_g = m_1 - m_0$. They only account for 1 or 2% of the total mass however, and can therefore be considered negligible.

The weight fraction of the resin is given by $M_m = 1 - M_F$ (assuming no voids are present).

To convert to a volume fraction, the formula given in Section 5.1.2(d) is used. The specific gravity or density of the resin ρ_m and of the fibre ρ_F are data provided by the supplier.

(d) Determination of carbon fibre volume fraction

Project European standard EN 2564 or ASTM D 3171-73, T57-608.

Since carbon fibres oxidize in air above 300 °C a burn-off method cannot be used. The resin is therefore chemically removed by a hot solution of concentrated sulphuric acid and 50-volume hydrogen peroxide. The difference in mass before and after the dissolution of the resin gives the mass of fibres contained in the sample.

A sample of laminate is conditioned for at least 16 hours under standard atmospheric conditions (23 °C, 50% RH). The sample is then weighed accurately (mass m_1) and placed in sulphuric acid at 160 °C in an Erlenmeyer flask. 50% W/V hydrogen peroxide is added until the solution becomes colourless. The solution is heated to 160 °C for a minimum of ten minutes, to ensure complete decomposition of the polymer matrix. After cooling, the contents of the Erlenmeyer flask are filtered using a sintered glass filter of known mass (m_2).

The carbon fibres which have not been dissolved are washed in distilled water and ketone. The filter and contents are dried in an oven at 120 °C, and cooled in a desiccator before being weighed: mass m_3.

$m_1 = $ initial mass of the test piece (fibres + resin).
$m_2 = $ mass of the sintered glass filter.
$m_3 = $ mass of the filter + carbon fibres.

Hence, the mass of carbon is $m_c = m_3 - m_2$. The weight ratio of carbon fibre in the laminate is:

$$M_F = \frac{m_3 - m_2}{m_1} \text{ (in %)}$$

the volume fraction is:

$$V_F = \frac{1}{\left[1 + \frac{m_1 - m_c}{m_c} \times \frac{\rho_F}{\rho_m}\right]}.$$

$\rho_F = $ density of fibre; $\rho_m = $ density of cured resin. This assumes that no voids are present.
Note:

- At least three to five test pieces should be used.
- The chemical attack must be prolonged until a constant mass is obtained.
- The volume of carbon fibres is V_F.
- The density ρ_F of the fibre must be given by the supplier.
- The density of the composite is determined with a pycnometer (see Section 5.1.3(b)).

(e) Determination of the density of Aramid fibres

The following is based on a method developed by Allred and Hall, Sandia Laboratories (New Mexico). Roasting cannot be used as the fibres would decompose. Using a solution of sulphuric acid and hydrogen peroxide would damage the aramid fibres.

A hot solution of concentrated nitric acid and dimethylsulphyde (DMSO) determines the chemical digestion of the resin. The aramid fibres are collected by filtering. They are not significantly attacked by this digestion and it is unnecessary to apply any correction for fibre loss (1% at most).

The aramid-resin composite sample is dried for 1 hour at 110 °C, and is then weighed: mass m_t. It is then immersed in the DMSO at 120 °C for three to seven hours until separation of the fibres occurs. The mixture is poured on to a sintered glass filter, mass m_1. The filtrate is treated with a DMSO solution at 120 °C for two to five hours to further dissolve the resin. Filtration can then be carried out. The remainder of the composite sample is treated with 10-volume HNO_3 and 90-volume DMSO for 30 minutes at 130 °C to remove any remaining resin. After filtration, the filament residue is rinsed over a sintered glass filter with a cold solution of $DMSO/HNO_3$. Finally, it is rinsed with distilled water.

The sintered glass dish containing the aramid fibres is then heated at 100 °C for an hour, cooled in a desiccator and weighed: mass m_2.

- Initial mass of the test piece: m_t.
- Mass of the sintered glass filter: m_1.
- Mass of the filter and aramid fibres: m_2.

Hence, the weight ratio of aramid fibres in the laminate is:

$$m_F = \frac{m_2 - m_1}{m_t} \text{ (in \%).}$$

The volume percentage of aramid fibres is:

$$V_F = m_F \times \frac{\rho_s}{\rho_F} \text{ (in \%).}$$

ρ_s is the density of the laminate. This is measured with a pycnometer (see Section 5.1.3). ρ_F is the density of the aramid fibre PRD149 (Du Pont de Nemours) or of Twaron (Enka-Akzo, The Netherlands). These values are given by the suppliers.

5.1.3 Density of the laminate

It is preferable to determine the density of the composite by submersion in a liquid. For this a pycnometer is used.

Standards: T51-201 AFNOR, NF ISO 1675.

(a) Density definition

The relative density of a solid of volume v is equal to the ratio of the mass of the solid to the mass of an equivalent volume, V, of water:

$$\rho = \frac{\text{mass of solid of volume } v}{\text{mass of the volume } V \text{ of water}}.$$

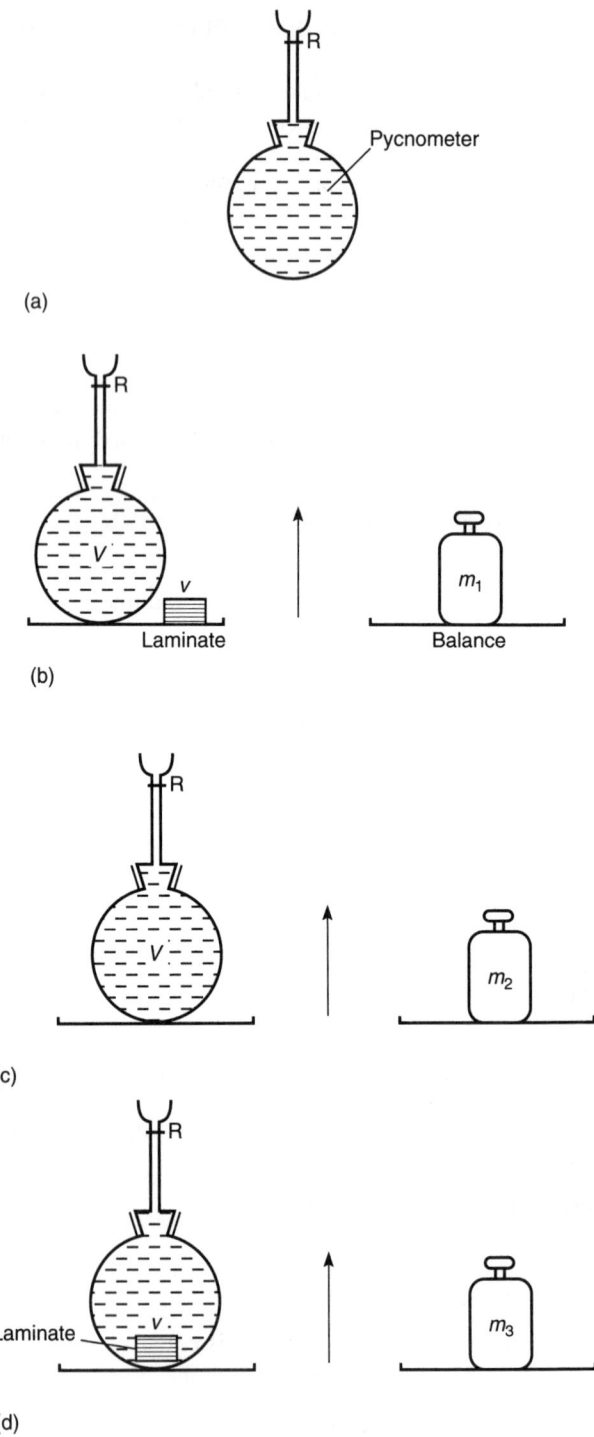

(a)

(b)

(c)

(d)

Figure 5.2

(b) Method using a pycnometer

A special bottle called a pycnometer is used. The hollow ground stopper contains a thin central tube bearing a mark R. Weigh the bottle, m_0. Weigh the bottle plus laminate, m_i. Fill the flask up to the mark (volume V) with water, making sure that no air bubbles are present. After wiping the outside of the flask carefully, weigh it, m_2.

Place the laminate sample in the bottle and weigh, m_3, after carefully wiping any water from the outside of the bottle. Then $\rho = (m_1 - m_2)/(m_1 - m_3)$. The steps are illustrated in Figure 5.2.

5.1.4 Void content

Hand-made laminating causes air bubbles to congregate between layers especially if successive layers have not been properly consolidated with a roller. The higher the mass per unit area of the woven fabric, the greater the likelihood of having bubbles or voids between the layers.

These voids may cause stress concentration and delamination. It is therefore necessary to check the porosity or void content of laminates.

(a) Method based on densities

The porosity of a test piece can be calculated from the following relation:

$$V_v = 100 - \left[\left(m_F \times \frac{\rho_s}{\rho_F} \right) + (100 - m_F) \times \frac{\rho_s}{\rho_m} \right].$$

In this relation, V_v is the void content, m_F is the fibre content expressed as a mass percentage, ρ_F is the density of the fibre, and ρ_m is the density of the matrix and ρ_s is the density of the laminate.

Example
Laminate of:

- E glass roving with a mass of $830\,g/m^2$.
- isophthalic polyester resin containing 1.2% of colloidal silica thixotropic agent.

$$m_F = 45\%$$
$$\rho_s = 1.5\,g/cm^3$$
$$\rho_F = 2.6\,g/cm^3$$
$$\rho_m = 1.2\,g/cm^3$$
$$V_v = 100 - \left(45 \times \frac{1.5}{2.6} \right) + (100 - 45) \times \frac{1.5}{1.2} ,$$

hence $V_v = 5.6\%$.

(b) Method based on microscopy

The void ratio can be obtained by making a quantitative observation of polished micrographic sections.

Manual method
Standard NFT 57-109, ISO 7822.

The sample of laminate composite is cut, mounted in and covered with a cold setting resin and carefully polished first with 1200 grain abrasive paper, and then with a series of very fine diamond pastes (8, 4, 1, 1/4 µ) or alumina. The process, illustrated in Figure 5.3, is the same as for polishing metallographic specimens. Polishing carbon fibre laminates is difficult because the fibres are brittle and may be partially pulled out, leaving apparent porosity.

A minimum of five sections must be prepared and examined to obtain a statistically satisfactory result. A rapid check at low magnification enables the porosity fraction V_v to be determined. Pores or voids are black against the light background. The method consists of superimposing grids containing N boxes (100 or more) on the micrograph. If $n \leqslant N$ boxes contain a void

$$V_v = \frac{n}{N}.$$

An eyepiece incorporating a grid can also be used.

In order to refine the result, the operation can be performed using higher magnifying power and finer grids containing more boxes. See Figure 5.4.

Automatic method
If a quantitative microscope is used, the result V_v can be obtained automatically. When the porosity ratio is low, this method is preferred, but it requires numerous sections to be examined if the relative error is to be small.

Standard NFT 57-109 provides a table which, depending on the relative

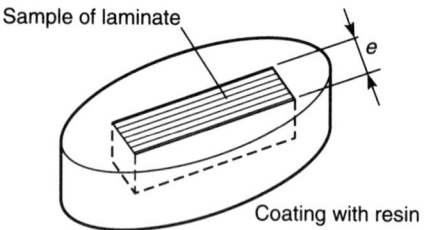

Sample of laminate

e

Coating with resin

Figure 5.3

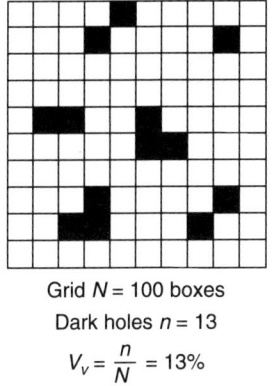

Grid N = 100 boxes

Dark holes n = 13

$$V_v = \frac{n}{N} = 13\%$$

Figure 5.4 Determining porosity.

error expected, e_v, and the porosity percentage V_v for the laminate, indicates the average numbers of boxes that must be examined, as shown in Figure 5.5.

The following relation can also be used:

$$e_v = 100 \sqrt{\frac{1 - V_v}{M + V_v}} \%.$$

A quantitative analysis microscope may be connected to image processing equipment. Using these techniques it has thus been established that the skins of sandwiches from which satellite structures are made contain less than 3% voids.

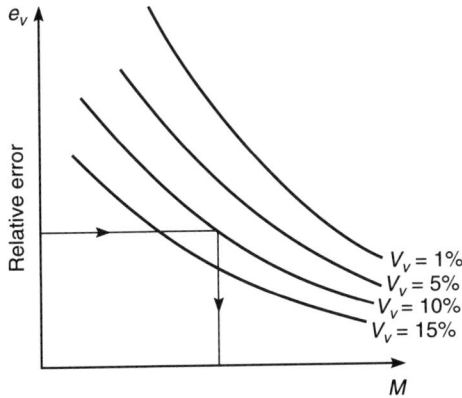

Figure 5.5 Average number of boxes to be examined.

The image processer shows:

- white, the epoxy matrix;
- grey, sections of carbon fibres;
- dark, the holes or voids which are to be detected.

The result is generally given to the nearest decimal point.

5.1.5 Curing control

This check is carried out after curing. The subject was dealt with in Section 4.2 'Controlling the curing cycle'. See Section 4.2.2(b) 'Checking for incomplete curing – tests carried out on moulded items'. The DSC method which allows the determination of the degree of cross-linking is the most effective. When cross-linking reaches 95%, the cure is considered satisfactory.

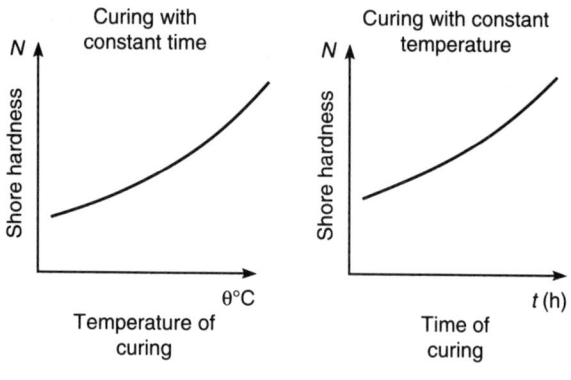

Figure 5.6

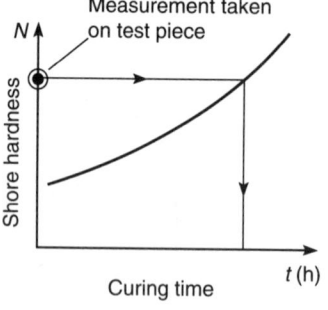

Figure 5.7

A rapid, empirical check of the state of cure, albeit inaccurate, is based on Shore or Barcol hardness – see standards NF ISO 868, or EN 59. This approach needs, however, properly cured reference samples for comparison. These must be made from the same resin and the same fibre and must contain the same volume loading of fibres as the test piece.

The curing characteristics must be well defined:

- increasing temperature for constant time;
- or constant temperature and increasing time.

Measurements must always be taken in the same way, for example perpendicular to the fibres; see Figure 5.6.

If the hardness number N is such that $N < 85$, use Shore hardness A. If it is such that $N > 85$, use Shore hardness D (see Figure 5.7).

5.1.6 Control of mechanical properties

These tests are exactly the same as those carried out to characterize composite materials. Their aim is to check that the characteristics assumed by the designer are correct. The easiest test to perform is the bending test and because of its simplicity and inexpensive nature, it is often used for laminates. We will begin by describing this test.

(a) Three-point bending test

Test piece
Standards are EN 63, NFT 57-105, ISO 178, L17-411 or EN 2562 projected.

The bending test, shown in Figures 5.8 and 5.9, is carried out on a rectangular cross-section beam cut from the composite. The beam is supported towards either end and loaded in the middle. The standard is EN 63.

The load F is applied slowly so that $y < 10$ mm/min.

The lower or under layers of the laminate are subjected to tension. It is possible to determine the following:

- failure strength in flexure σ_F;
- the apparent bending modulus, E_F;
- the maximum bending displacement, y_m, at ultimate stress, given by the displacement under the loading point.

Maximum tensile stress σ_m
The maximum tensile stress σ_m occurs in the middle of the bar. The bending stress is $\sigma = M_t e / 2I$ where M_t is the bending moment for load F and I is the moment of inertia of the beam section.

$$M_{t\,max} = \frac{F}{2} \times \frac{D}{2} = \frac{FD}{4}, \quad I = \frac{1}{12} be^3$$

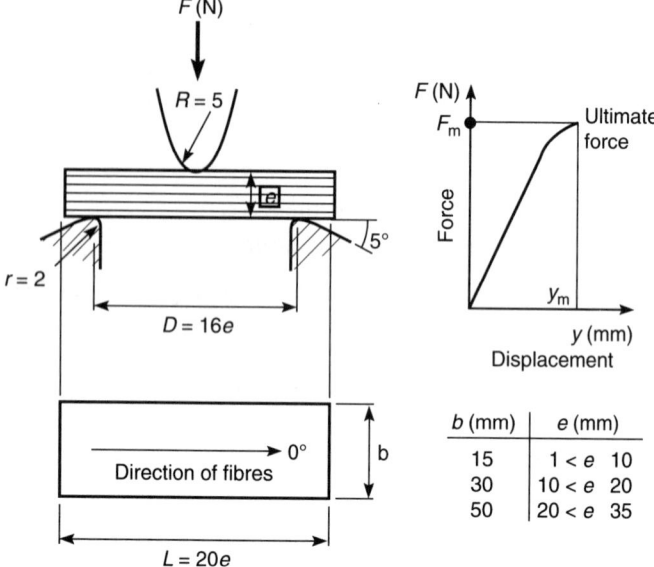

b (mm)	e (mm)
15	$1 < e$ 10
30	$10 < e$ 20
50	$20 < e$ 35

Figure 5.8 Flexural test (after EN 63).

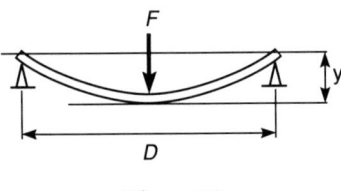

Figure 5.9

for a rectangular section where D is the length of the beam, see Figure 5.10. Hence,

$$\sigma_{max} = \sigma_m = \frac{3FD}{2be^3}$$

with F in newtons and b, D, e in mm. The result is usually quoted in MPa. Frequently the flexural compressive stress is taken to be similar to the flexural tensile stress.

Stiffness modulus E_F
A point A is chosen on the straight part of the force–deflection curve shown

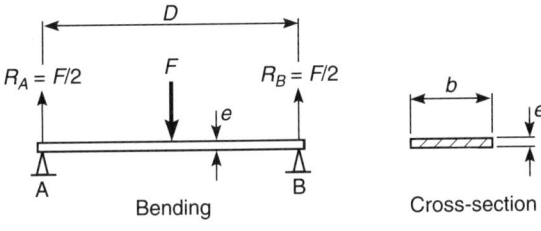

Figure 5.10

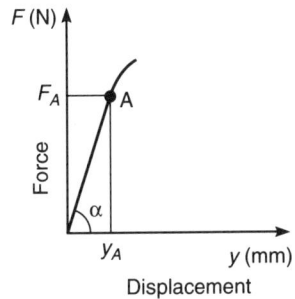

Figure 5.11 Determining modulus E_m through bending

in Figure 5.11 (in the elastic region) $\tan \alpha = F/y_A$, and we have:

$$E = \frac{D^3}{4be^3} \tan \alpha$$

Note this assumes that D/e is sufficiently large that shear deflection is negligible. The requisite value of D/e depends on the reinforcement. 16:1 is a suitable value for GRP, 40:1 for CFRP.

Notes
For quality control, specimens should be cut parallel to the principal fibre axis. If test pieces are too thick and D/e too small, failure may not occur under tension/compression, but by shearing or delamination; see Figure 5.12.

Failures by shear or delamination should be discounted. The value of D/e must be increased to avoid this behaviour. The test requires several test pieces to be used so that the result may be of statistical significance. The maximum stress within the test piece only occurs in a limited volume, theoretically on the underside of the test piece and at its centre, i.e. on a line

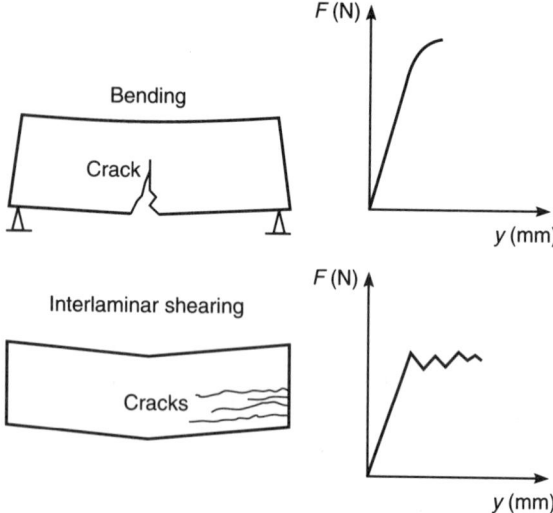

Figure 5.12

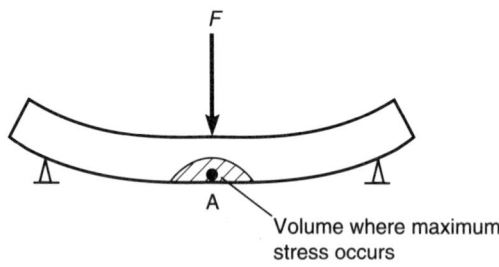

Figure 5.13

A seen in Figure 5.13. Given the fact that composite materials are varied in composition, the values of strength σ_F, and flexural modulus E_F will be somewhat scattered.

In this test the spread in results is higher than in the tensile test. However, the test subjects the material to different stresses: tensile on one side; compression on the other shear, etc., and produces results characterized by a high degree of variation. The test is often adapted to quality control requirements because it is a very simple and practical one both from the point of view of the preparation of the test pieces and from the method of test.

(b) Tensile test

European standard EN 61, AFNOR NFT 57-101, DIN 53455/392-457, ASTM D 3039, BS 2782 part 10 method 1003, ISO 3268, L17-410, PrEN 2561.

This test may seem simple enough, but in fact it requires care in cutting the test piece, as the tools used may leave marks which may later cause failure by delamination. Moreover, in the case of very stiff and strong composites (carbon, aramid), especially unidirectional ones, it is necessary to add end reinforcements in order to avoid crushing the test piece where it is clamped at the ends. If this is not done, failure may initiate in the crushed region.

Test piece
As a precaution when conducting a quality control test, the test piece should always be fitted with reinforcements at both ends. These should be glued on with a strong epoxy adhesive or riveted in accordance with type III shown by ISO standard, as in Figures 5.14 and 5.15. Five results should be obtained which show tensile failure in the gauge length. Consequently, at least five test pieces should be cut with a diamond saw or high-pressure water jet from the component or an extra length of material produced at the same time as the main component.

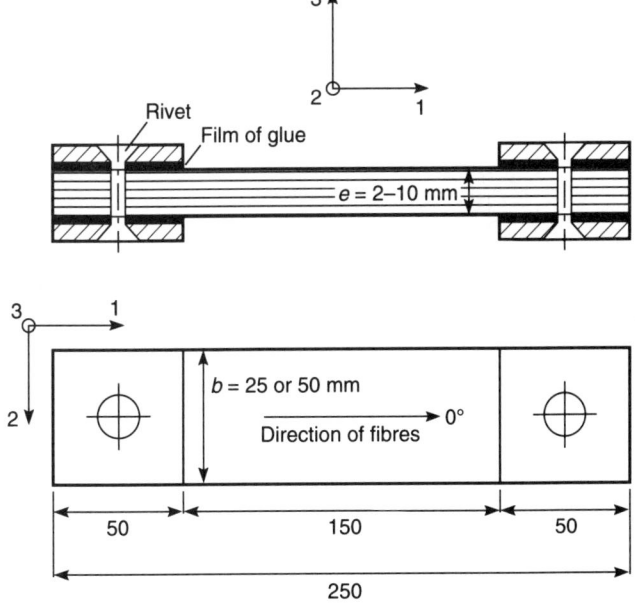

Figure 5.14 ISO Type III test piece.

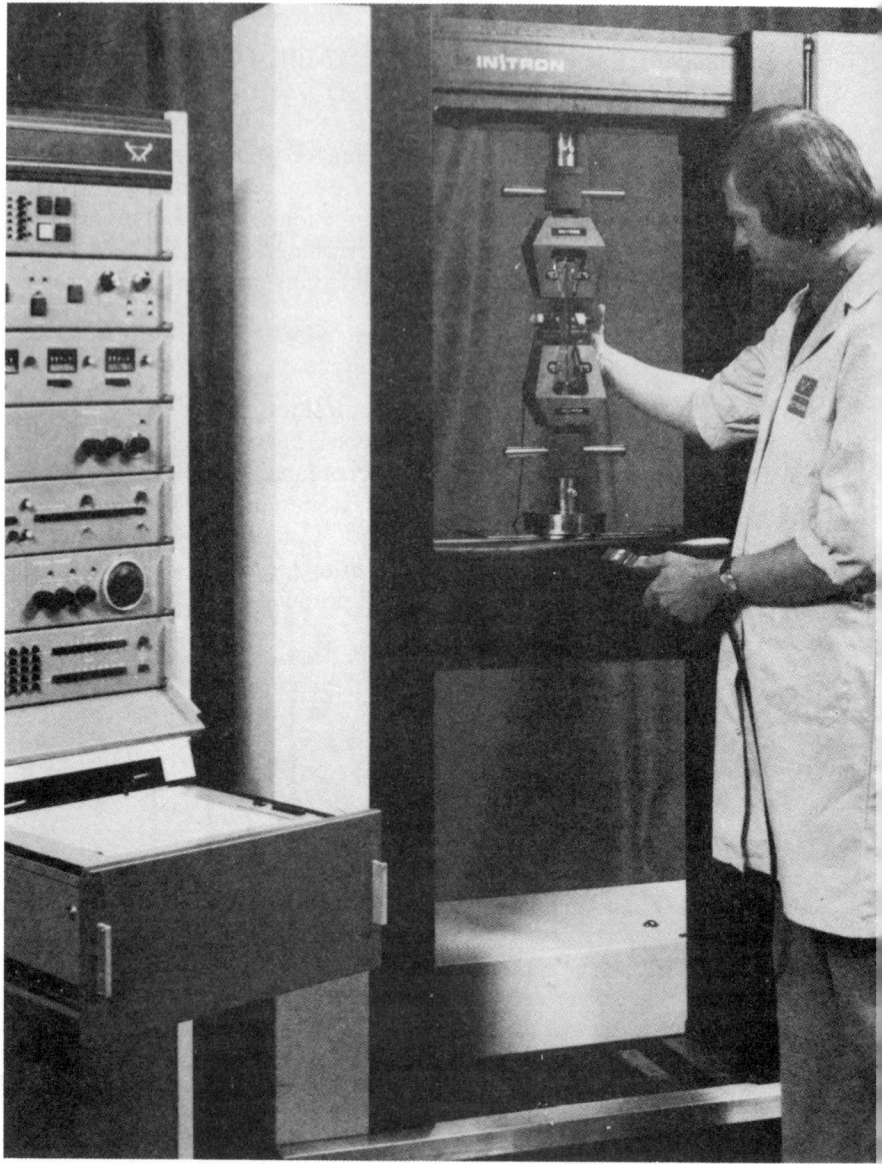

Figure 5.15

Notice that type III test pieces are larger than the others and it is sometimes impossible to cut them from the main work piece. In this case, shorter test pieces may be used.

Note: With UD laminates the thickness e is usually limited to 2 mm and the width b to 12.5 mm otherwise the required break load might exceed the tensile machine's capacity.

Straining speed
$V \leqslant 5\,\text{mm/min}$ for a routine check where only maximum resistance load is measured (no extensometer used), or $V \leqslant 2\,\text{mm/min}$ if one wishes to determine strain ($A\%$) at the break load and the stiffness modulus E. In this case the test piece is fitted with an extensometer.

Results

Routine checks
The load/displacement curve ($F, \Delta l$) may or may not be obtained in full, as in Figure 5.16.

$$\sigma_m = \frac{F_m}{be} \ (\text{MPa})$$

σ_m is the stress (MPa), F_m is the maximum load (N), b is the width of the test piece in mm, e is the thickness of the test piece in mm.
  Full scale test (rarely carried out). The load–strain curve (F, ε) is obtained,

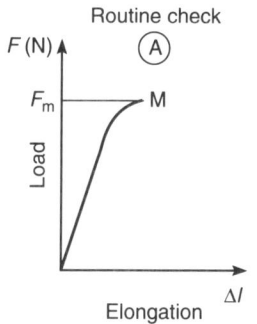

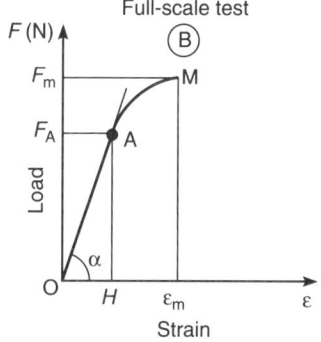

Figure 5.16

where

$$\varepsilon = \frac{\Delta l}{l_0}.$$

The initial part of the curve OA is usually linear.

$$\sigma_m = \frac{F_m}{b \times e} \text{ ultimate strength.}$$

At point A $\sigma_A = \sigma_e = \dfrac{F_A}{b \times e}$ yield strength.

$$E = \frac{\sigma_A}{\varepsilon_A} \text{ elastic modulus.}$$

$A\% = \varepsilon$, ultimate strain at failure, % (at point M).

Notes

The results from test pieces that move or break in the clamps or fail at less than 10 mm from the clamps must be rejected. For unidirectional laminates it is difficult to avoid failures close to the end reinforcements. Fortunately, this does not produce a noticeable effect on the results. In order to minimize the number of unacceptable tests the clamps should operate on the entire width of the reinforcement and the clamping procedure should be specified in the standard. Generally speaking, ultimate failure stress in flexure is higher than that in tension. The reason for this is that stress is not even in flexural tests, but varies through the thickness.

The results obtained when testing composite materials are dependent on the dimensions of test pieces. The volume stressed in tension in a test piece subjected to a flexural stressing is triangular in shape and is smaller in volume than in a tensile test. The likelihood of there being defects in the former case is therefore less, hence the average strength is higher. See Figure 5.17.

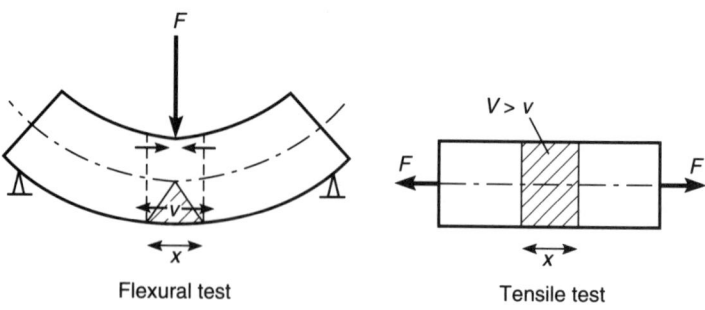

Flexural test   Tensile test

Figure 5.17

(c) Interlaminar shear tests

The problem
One of the weaknesses of laminates is shearing between the reinforcement layers.

Shearing occurs either at resin–fibre interfaces or within a layer of resin owing to a defect such as a pore or void, as shown in Figures 5.18 and 5.19.

Interlaminar shear test
The simplest test consists of loading a short beam in three-point loading, see standards NF ISO 4585 or ASTM D 2344. The test beam, whose thickness is e, is placed on supports $L = 5e$ apart. The load is applied on the middle of the test piece at a very low speed ($V \leqslant 1$ mm/min). The test piece must have a low ratio $l/e = b$ so that it does not fail by shearing when loaded. See Figure 5.20.

In general, the width b of the test is 10 mm. The interlaminar separation or cracking which occurs under the shear stress τ, parallel to the layers of the laminate, is characterized by a drop in load in the force–displacement curve and an audible noise.

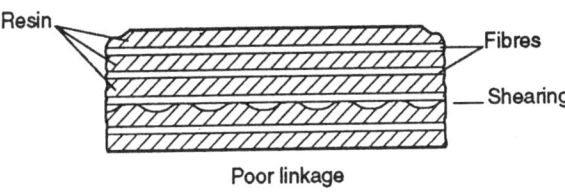

Figure 5.18 Shearing at resin–fibre interface.

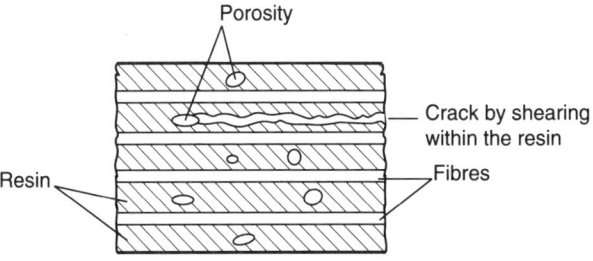

Figure 5.19 Cracks initiated by porosities due to low fracture toughness (G_{1c}) of the thermoset resin.

The interlaminar shear strength (ILSS)

$$\tau_{13} = \frac{3}{4} \frac{F_m}{be} \text{ (MPa)}.$$

Results are generally characterized by a high degree of dispersion. It is therefore necessary to use at least ten test pieces. See Figure 5.21.

Interlaminar shearing test of UD laminates stressed in three-point loading
The appropriate standards are L17-412, PrEN 2563 and NFEN 2377. These standards are for quality control tests carried out on unidirectional carbon fibre laminates, with the length of the test piece being parallel to the long fibre axis.

The deformation rate must be slow: $V \simeq 1\,\text{mm/min}$. The distance between the supports must be small to give a span-to-depth ratio of approximately 5:1. This ensures that flexural stress is negligible; see Figure 5.22.

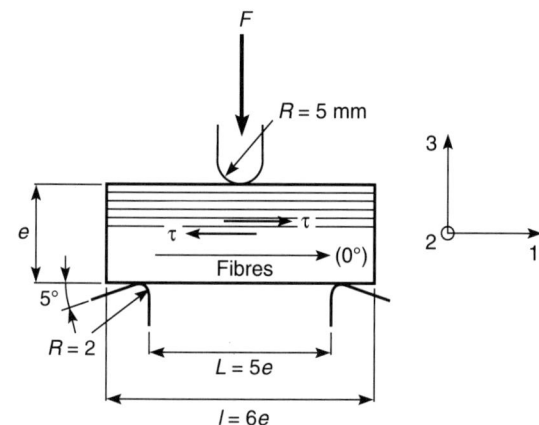

Figure 5.20 Bending shear test (short beam).

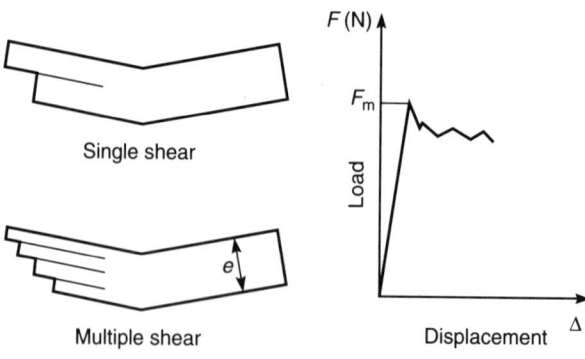

Figure 5.21

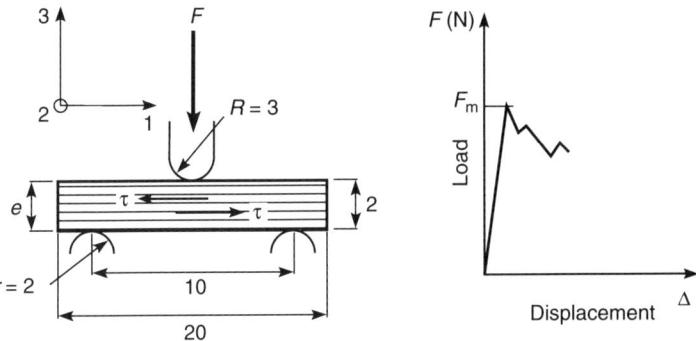

Figure 5.22

The width b of the test piece is 10 mm. The shearing or apparent interlaminar stress is given by:

$$\tau_{13} = \frac{3}{4} \frac{F_m}{be}$$

where F_m is the load at failure.

Shear testing under tensile loading

The surface under the central loading point is subjected to compression and the composite may well buckle if the laminate is made of a material such as aramid fabric, which is not very resistant to compression loading (Figure 5.23). In this case it is better to load the specimen in tension; see standard NFT 57-554.

This test can be extended to glass and carbon fibre laminates. The flat test piece, of thickness e, contains two notches, cut with a diamond disc. The notch depth is $e/2$. The difficulty with this test is in machining the notches to an accurate depth; see Figure 5.24.

Tension is applied slowly on the test piece along its axis (at a rate $V \leqslant 5$ mm/min) until failure occurs in the plane over the area, ABCD.

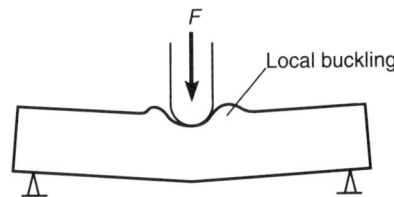

Figure 5.23 Shearing test on composite aramid/resin.

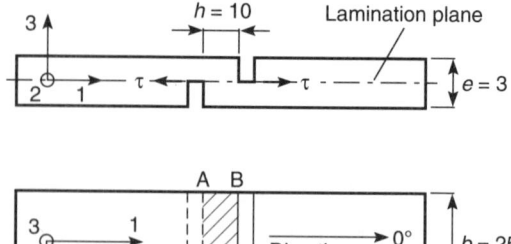

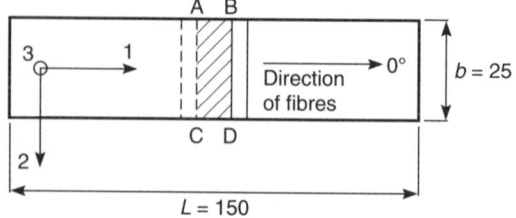

Figure 5.24 Interlaminar shear test under traction.

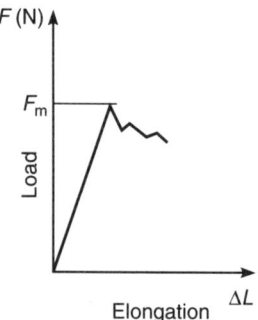

Figure 5.25

The shearing stress is:

$$\tau = \frac{F_m}{b \times h} \ (MPa)$$

The final result, Figure 5.25, should be the average of the minimum of five replicates.

Further shear tests
New tests have been introduced, but they are for characterizing composite materials rather than for quality control.
    Among these tests are:

• Iosipescu (1960) which has been used in the USA for several years;
• rail shear test.

The Iosipescu test is speedy and of general applicability. Machining the test piece is difficult, however. The test requires the use of a specimen with 90° notches as shown in Figures 5.26 and 5.27. The test piece is clamped in an auxiliary device which itself is fitted on to a tension–compression testing machine. When loaded a shear stress is generated in the middle of the test piece. The dimensions of the test piece can be reduced if required. The line AB shows the shear plane.

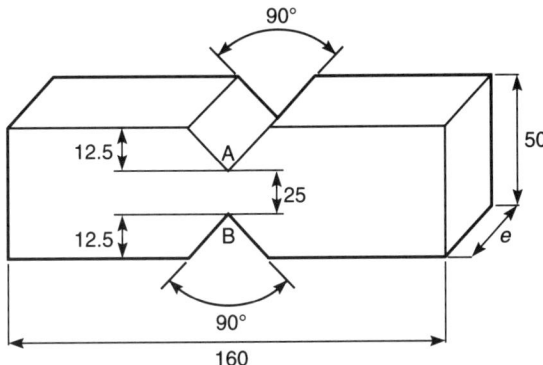

Figure 5.26 Iosipescu test piece.

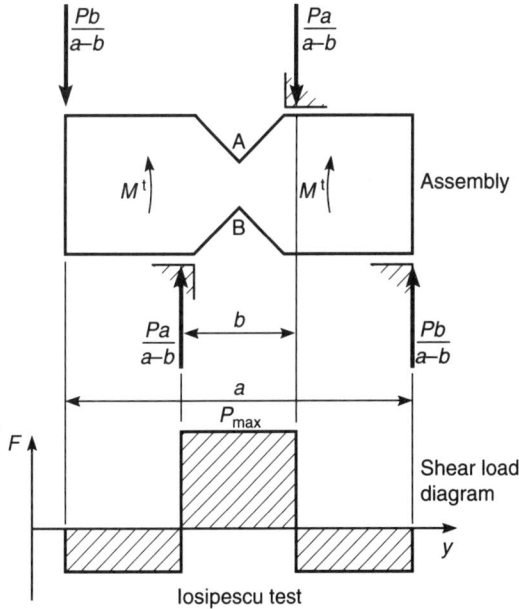

Figure 5.27

Depending on the arrangement of layers relative to the line AB, we have the different shear stresses τ_{12} or τ_{31} or τ_{23} shown in Figures 5.28 and 5.29.

In the rail shear test shown in Figure 5.30, a laminate sheet of thickness e and length l is bolted on both sides on to fixed rails. The jig plus specimen is loaded until shearing occurs. The sections torn are: $S = 2\,AB = 2 \times l \times e$, and the shear stress is

$$\tau = \frac{F_{max}}{2l \times e}.$$

The layers of fabric are generally placed parallel to the plane (1,2). The torn sections are planes (1,3), as shown in Figure 5.31; hence the measurement of intralaminar shear stress τ_{12}. This assembly is not useful for quality control.

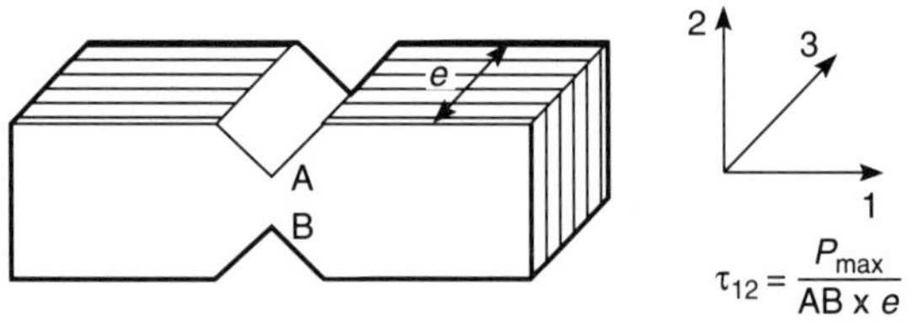

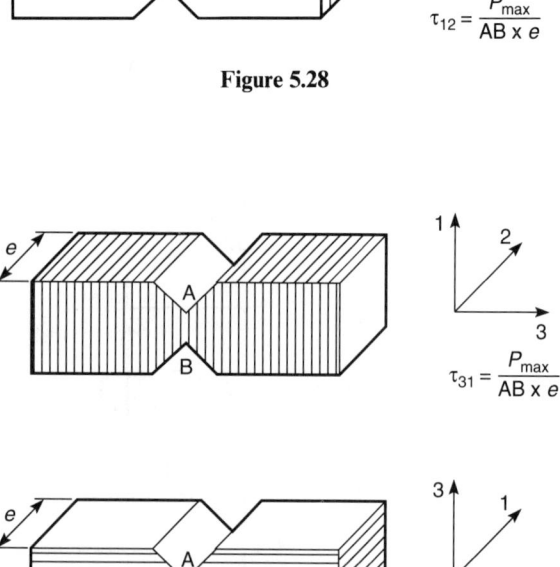

Figure 5.28

$$\tau_{12} = \frac{P_{max}}{AB \times e}$$

$$\tau_{31} = \frac{P_{max}}{AB \times e}$$

$$\tau_{23} = \frac{P_{max}}{AB \times e}$$

Figure 5.29

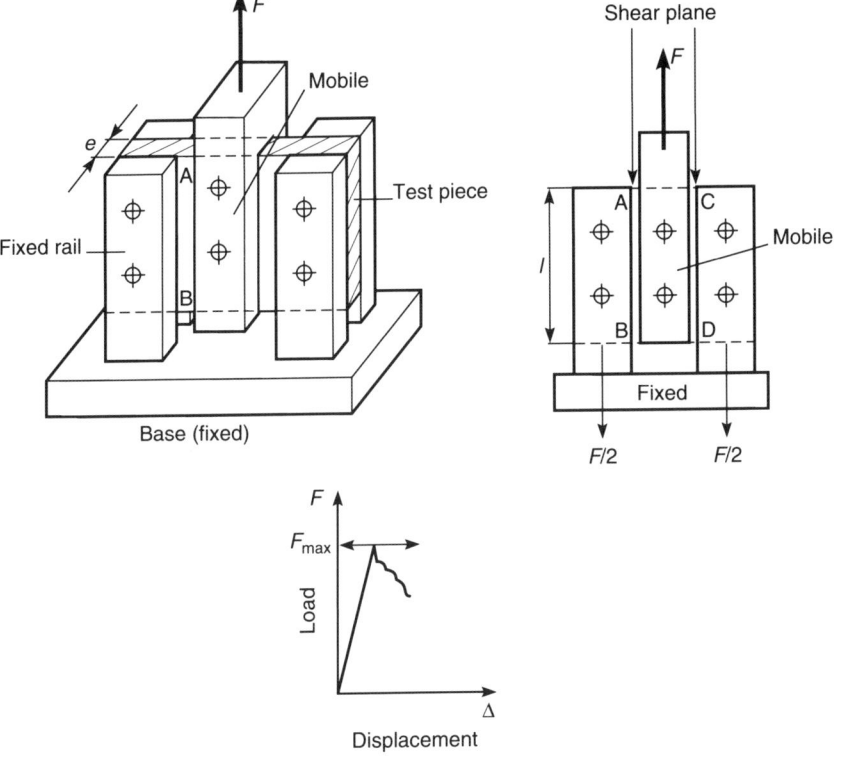

Figure 5.30 Rail shear test.

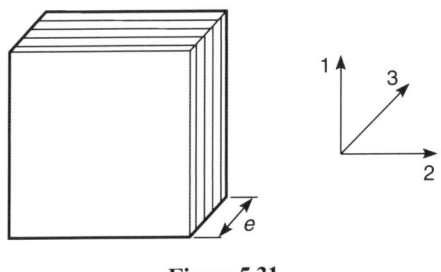

Figure 5.31

(d) Charpy impact strength test

See standards NFT 57-108, ISO 179 and DIN 53453.

This test is suitable for production checks, but requires the use of at least ten test pieces owing to a large spread in the results.

The test piece

The specimen is a parallel-sided bar of dimensions (x, y, L) shown in Figure 5.32, which is placed resting on two supports a distance D apart and struck in the centre. The test piece is notched. y is the thickness of the test piece. If it is too thick, it must be machined down to $y = 10$ mm. $x = 10–15$ mm depending on the thickness of the fabric. $L = D + 5x$.

There are two possible test piece configurations:

$$\frac{D}{x} = 20, \quad \text{type I} \quad \text{or} \quad \frac{D}{x} = 6, \quad \text{type II}$$

Assembly

The specimen is impacted perpendicular to the plane of the laminated layers and if large enough failure occurs either by tension in test pieces of type I or by shearing in test pieces of type II. See Figure 5.33.

Results

The Charpy impact strength is the energy absorbed up to the failure point by the test piece in the volume between supports:

$$K = \frac{A}{x \times y \times D} = 10^9$$

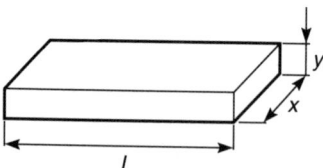

Figure 5.32

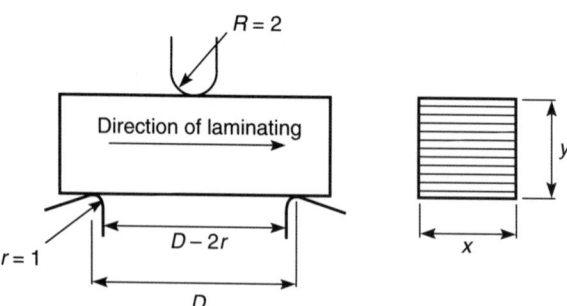

Figure 5.33

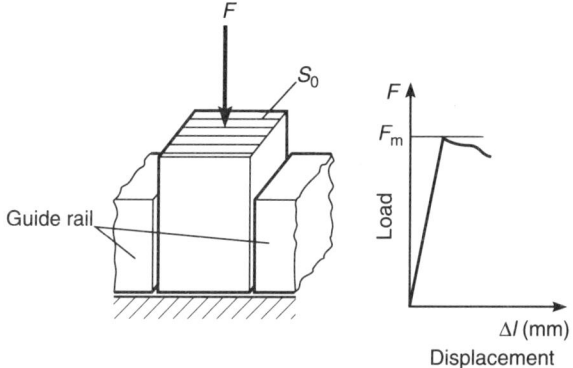

Figure 5.34

where K is the impact strength, in J/m^3, A is the energy absorbed, in J, x, y and D are in mm.

(e) Compression test

See standards NF ISO 8515, DIN 53-454.

Carrying out this type of test on composite materials presents quite a number of difficulties, which is why it is rarely used to characterize these materials. It is rarely used in manufacturing checks.

It is even more difficult to carry out compression tests on aramid laminates (Kevlar, Twaron, fibre based) because microbuckling occurs at a very low compressive stress in this type of laminate.

A block-shaped test piece of section S_0 is guided in order to avoid buckling under the load F applied to the top, as shown in Figure 5.34. The load F is parallel to the layers of the laminate. Unfortunately, the load crushes the resin coated fibres at the ends of the test piece. In a quality control test, only the load displacement reading is noted.

The stress or maximum resistance under load is

$$R_c = \frac{F_M}{S_0}.$$

5.1.7 A note on the statistics used in quality control checks

(a) Gaussian curve

Let us consider N objects, items or events, and study the statistical distribution of one of their physical characteristics, for instance, size. The arithmetic

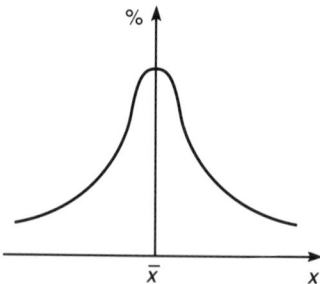

Figure 5.35 Gaussian curve.

average of the characteristic x (size) is:

$$\bar{x} = \sum_{i=1}^{i=N} \frac{x_i}{N}$$

where N is the number of items studied and x is the size of each item.

Plot the results: on the x-axis the value x; on the y-axis the percentage of items of size x. We get a symmetrical distribution about the average value $\bar{x}$, in the shape of a bell, the so-called Gaussian distribution shown in Figure 5.35.

(b) Parameters on which a statistical distribution is based

Two parameters are essential:

- the average, $\bar{x}$;
- the standard deviation, s, given by the following relations:

$$s = \sqrt{\frac{\sum_{i=1}^{i=N} (x_i - \bar{x})^2}{N-1}} = \sqrt{\frac{\sum_{i=1}^{i=N} x_i^2 - \frac{1}{N}\left(\sum_{1}^{N} x_i\right)^2}{N-1}}.$$

where x_i is the value of one of the measurements taken, N is the total number of measurements taken and $\bar{x}$ is the arithmetic average obtained from N measurements. Each item is assumed to be distinct and independent. All items are selected at random.

95% confidence interval

An interval containing 95% of the items is determined about the average $\bar{x}$; this is known as the 95% confidence interval. Ninety-five percent of the results obtained from analysing the set of items are located between $(\bar{x} - 2s)$ and $(\bar{x} + 2s)$ as shown in Figure 5.36.

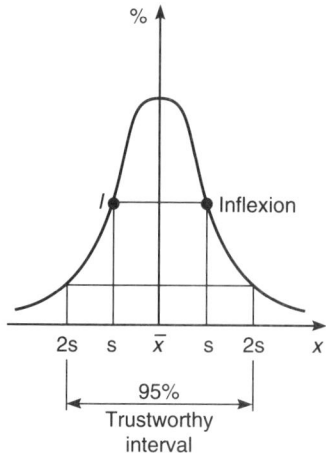

Figure 5.36 The 95% confidence interval.

If I is the inflexion point of the Gaussian curve, the standard deviation corresponds to the distance D between the average $\bar{x}$ and the point I. The 95% bilateral confidence interval is defined by:

$$\bar{x} - \frac{ts}{\sqrt{N}} \leqslant \bar{x} \leqslant \bar{x} + \frac{ts}{\sqrt{N}},$$

Table 5.1 Student–Fisher values

N	t	N	t
2	8.985	19	0.482
3	2.485	20	0.468
4	1.591	21	0.455
5	1.241	22	0.444
6	1.049	23	0.432
7	0.926	24	0.422
8	0.836	25	0.413
9	0.769	26	0.404
10	0.715	27	0.396
11	0.672	28	0.388
12	0.635	29	0.380
13	0.604	30	0.373
14	0.577	40	0.319
15	0.554	60	0.259
16	0.533	80	0.223
17	0.514	120	0.181
18	0.497	250	0.125

where $(t)/\sqrt{N}$ is given in the Table 5.1.

(c) Application to quality control tests carried out on a component

Sampling

Owing to price considerations, only a small number of samples can be selected for physical, chemical and mechanical quality control tests. Typically $N = 5$–10 samples or test pieces. All results that are obviously in error for experimental or technical reasons must be rejected before the statistical analysis is carried out.

After these deviant results have been rejected, those retained consist of N measurements, values $x_i = x_1, x_2, x_3, \ldots, x_n$, some of which may be equal.

Statistical formulae to be used

$$\text{Average } \bar{x} = \frac{\sum_1^N x_i}{N};$$

$$\text{standard deviation } s = \left[\frac{\sum_1^N (x_i - \bar{x})^2}{N - 1} \right]^{1/2}.$$

The 95% confidence limits are:

$$\bar{x} - \frac{t}{\sqrt{N}} \times s \leqslant \text{average } \bar{x} \leqslant \bar{x} + \frac{t}{\sqrt{N}} \times s.$$

5.2 NON-DESTRUCTIVE TESTING (NDT) ON A PRODUCT

5.2.1 Specific characteristics of tests conducted on composites

NDT is used to detect defects either in new items, as a check on new items, or in items already in service, as maintenance checks. In both cases the item is not damaged in any way by the test procedure. The main aim is to detect poor compaction leading to resin-rich areas or low fibre loading, bad surface appearance or internal defects such as:

- lack of reinforcement;
- porosity;
- poor fibre–matrix bonding (delamination);
- cracks of all sizes;
- failure of the adhesive bond between components;
- inclusions.

Note that the notion of defect or discontinuity in a composite material is enhanced since these materials are characterized by their heterogeneous nature.

Any defect that is detected must be determined:

- in area or volume;
- by its location within the structure in relation to the layers of reinforcement;
- by its relation to the mechanical stresses to which the material may be subjected.

A defect of a constant type and size may be unimportant in one case, but serious in another. Thus a series of air bubbles between two layers may be harmful if the material is subjected to a shear or flexural stress but the same defect occurring in a different item subjected to tension or compression would be of no consequence.

Tremendous advances have been made in the field of NDT in the last few years. Formerly the philosophy was to check for the absence of defects. Currently the trend is towards differentiating between those that are harmful to the structure and those that are not. Considerable effort is being made to automate the methods used in order to cut down costs and reduce the risk of human error.

Universal criteria for assessing the seriousness or otherwise of defects cannot be defined in the absence of a complete knowledge of the artefact and the way it will be employed.

5.2.2 List of methods used

Owing to the fact that the aerospace industry requires fail-proof parts, non-destructive test methods were first introduced by companies working in this field. Individual companies have their own acceptance/rejection criteria but do not publicize these.

The techniques used are complementary. For example X-rays and ultra-sonics (US) or IR thermography and ultrasonics.

In principle the methods used are very often similar to those recommended for homogeneous and metal objects. However, their application to hetero-geneous laminates requires further laboratory study. This is the reason why there are few if any standards to date for NDT applications. Research establishments and commercial committees are trying to establish standards in this area.

Methods fall into two main categories: qualitative and quantitative:

Qualitative tests:

- visual observation and dye penetration;
- photometry;
- ultrasonics (echography, US);
- X-rays, γ-rays and neutron radiography;
- IR thermography;
- vibrothermography;

- laser holographic interferometry;
- eddy currents;
- the analysis of light impacts (tapping).

Quantitative tests
There are a number of qualitative methods, many similar to those listed above. However, they are used in a different way combined with image analysis:

- ultrasonics (echography, US);
- X-rays, γ-rays and neutron radiography;
- IR thermography;
- laser interferometry;
- eddy currents;
- microwaves.

Other methods that also form the basis of quantitative tests are:

- acoustic emission;
- stimulated stress wave emission.

A different method of classification can be used: methods based on examining load-free parts, and those for examining stressed components.

Methods based on the observation of load-free parts
X-rays:

- method based on absorption, X-ray or neutron radiography;
- method based on X-ray diffusion and the Compton effect;
- tomography.

Visual inspection of parts exposed to light:

- direct eye observation;
- photometry;
- dye penetration test in daylight;
- dye penetration test in UV light.

IR thermography by transmission or absorption.
  Eddy currents, for conductive composites, such as those made of carbon fibre.
  Ultrasonics: method based on mapping the response and measuring the speed of the signal.

Methods based on the use of mechanical stress

- vibrothermography;
- acoustic emission;

- holographic interferometry;
- tapping;
- mechanical or thermal deformation, and optical metrology.

5.2.3 Visual inspection – dye penetration test

(a) Visible defects

Standard NFT 57-100 defines a certain number of defects affecting the surface appearance of the part which can be detected by the naked eye including: bubbles close to the surface, blisters, resin spew, cracks, craters or lack of reinforcement, surface flaking, local excess of reinforcement, and the presence on the surface of non-wetted fibres.

The following defects can easily be detected by carrying out a dye penetration test if the defects affect the surface: cracks in the gel coat, flaking, cracks, surface delamination due to machining (drilling, milling, cutting, etc.).

(b) Dye penetration test: principle

When certain dyes are applied, even the smallest microcrack connecting with the surface can be detected. The dye may be coloured or fluorescent. After rinsing the part in water and drying it, a white powder is applied which absorbs the dye. This process reveals dyed or fluorescent patches which can be detected in UV light.

(c) Special products for composite materials

Special dye penetrating agents have been developed for composite laminates. A typical agent is non-flammable and does not contain any organic solvents that might attack the resin matrix. In addition, the dye is biodegradable and can be rinsed off in water. Under UV light the dye emits a bright yellow–green fluorescent light. A dry powder is used to intensify the effect of the penetrating agent and reduce the leakage around defects which have absorbed the dye.

(d) Method

1. Clean the surface of the composite by using an alkaline degreasing agent or trichloroethane. Rinse the composite in water, and dry it in an oven (40 °C) for one or two 2 hours.
2. Apply the penetrating dye by spraying, brushing or immersing the composite in it. Allow at least 20 minutes for the agent to penetrate.
3. Rinse in running water or by using a low-pressure (2 bars) water–air jet for a limited time (30–60 seconds) in order to avoid flushing the agent out of minor defects. It helps to view the component in UV during this operation to check on the lamination of excess dye.

4. Thoroughly dry in an oven (40 °C).
5. Spray evenly with a jet of powder. Smaller parts can be immersed in a container filled with a suspension of powder.
6. Leave the agent to act for 15 minutes.
7. Inspect in UV light in a darkroom (direct light or sloped light).

Voids or porosity appear as bright spots, cracks as bright lines, against a black background. A scanner or image processor can be used to assess fluorescence automatically.

Detected defects: 20 µm in UV light, 40 µm in normal light.

5.2.4 Visual inspection – checking for opacity

With thin, glass fibre laminates, free of colouring pigments, it is possible to check the quality by viewing the material with the naked eye in ordinary or artificial light. If the composite has been well impregnated with resin and the fibres properly wetted, it will be translucent. Defects such as delamination or voids will show up as opacity.

The following illumination can be used:

• a powerful light source like neon or halogen, for instance – the laminate will act as a defuser;
• a beam of monochromatic red light.

The intensity of the transmitter light can be evaluated visually by an expert or measured with a photodiode.

Systematically scanning the surface of the composite with an image processor allows a defect map of the composite to be built up.

This method can only be used with very thin components of glass fibre reinforced material.

5.2.5 Ultrasonic tests (US)

(a) General

Low-frequency ultrasonic (0.2 to 15 MHz) can propagate through composites. The anisotropic, heterogeneous nature of the material causes alteration and dispersion of the beam. From a theoretical point of view, propagation is a complex phenomenon because of the two-phase nature of the system (fibre and resin). Structural defects interfere with the propagation of US and partially reflect the sound waves. Hence two methods have been developed based on reflection and transmission of the signal. At best lamination defects measuring 0.2 mm can be detected.

The presence of voids reduces the transmission of US, and the method used can be used to quantify void content. The transmission speed of US

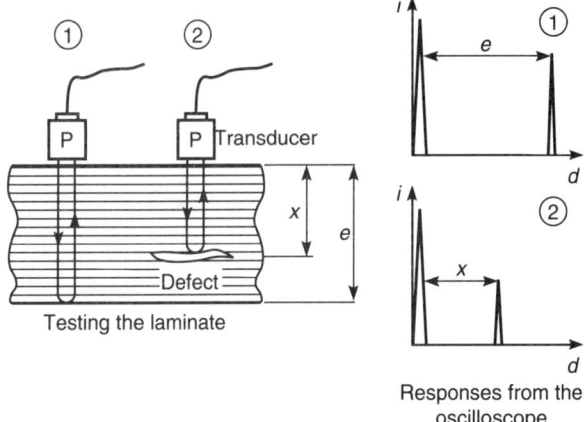

Figure 5.37 US reflection test.

·increases with increasing fibre loading and the method can be used to determine the fibre loading of a unidirectional composite.

The US method is complementary to X-ray or IR thermography, and is suitable for the detection of internal delamination.

(b) Reflection test

This method, is illustrated in Figure 5.37, is used to check components which can only be viewed from one side (very large pieces or containers). It would be possible to measure the thickness of composites with a resolution of 1/10 mm. Coupling between the transducer P and the laminate piece is usually facilitated with a liquid (water, oil, etc.). A water-jet transducer is sometimes used. If a coupling liquid cannot be used because it might damage the composite, special transducers fitted with rollers made of silicone elastomer which allows the passage of ultrasonic signals are used.

(c) Transmission test

This method is used for small or medium-sized components for which there is access to both sides; see Figure 5.38. Other artefacts may be immersed in a pool of liquid.

The equipment needed comprises:

- a US emitter–receiver;
- a position bench which allows the transducer to scan systematically the surface of the composite at a suitable rate;

*** S T C A N ***
defaut
Date : 20 03 87
Divers : NUMERO 5
Palpeur : 2.25 M
Col.d'eau : 45

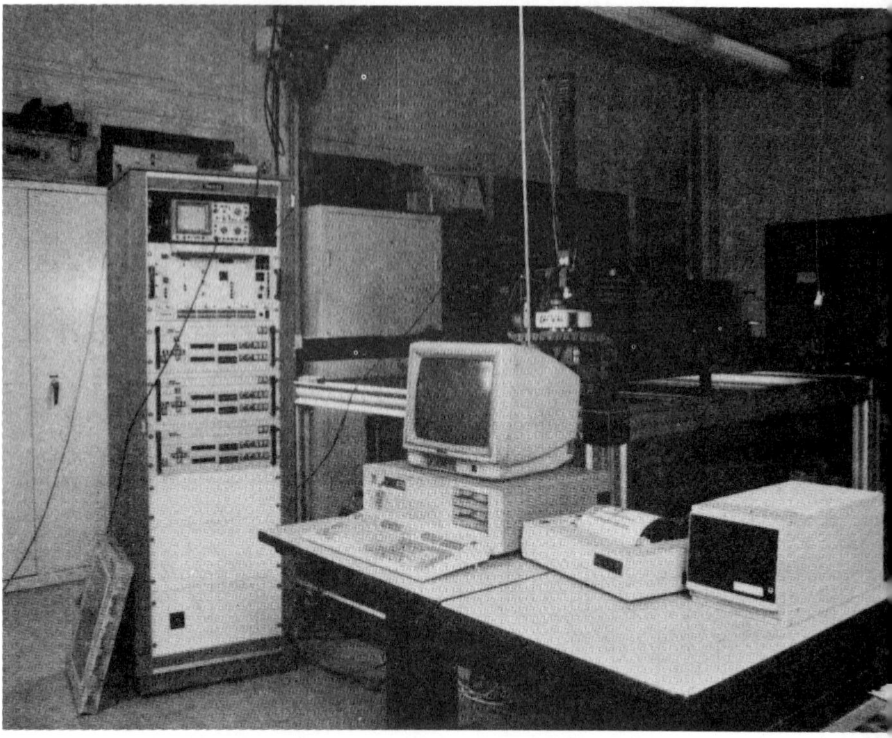

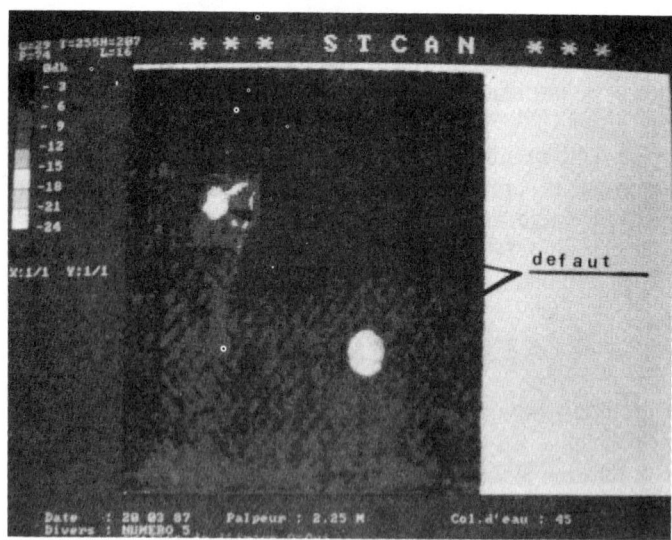

Figure 5.38

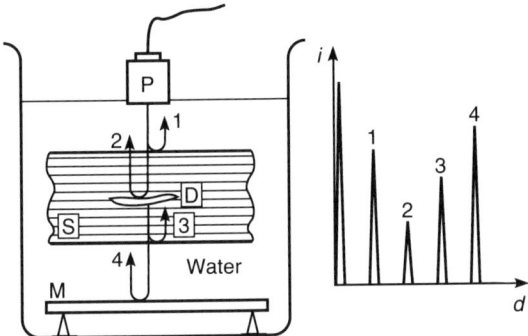

d Distance from the transducer
P Transducer
S Laminate
D Defect
M Reflecting mirror

Figure 5.39 US transmission test.

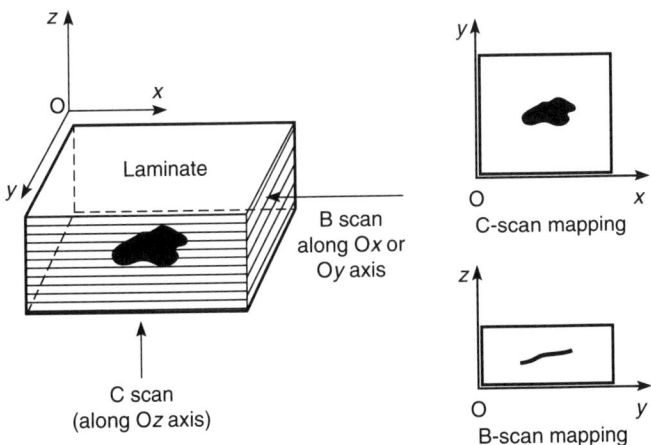

Figure 5.40 B- and C-scan mappings (principle).

- a pool filled with the coupling liquid (water and corrosion inhibitor);
- a digital oscilloscope;
- a microcomputer to monitor the test.

The test is illustrated in Figure 5.39.

The signal from the transducer is reflected by:

- the upper side of the material 1;
- the defect to be detected 2;
- the lower side of the laminate 3;
- the reflector M: a glass or aluminium sheet.

It then traverses the material again back to the emitter–receiver transducer.

The technique can be used to map the surface of the artefact to produce a B-scan or C-scan depending on whether the laminate is scanned from the side or from above. C-scanning is used for the qualitative detection of delamination or porosity as shown in Figure 5.40.

It is very difficult to assess the potential harmful effects of the defects detected (i.e. to determine the exact extent and depth of the crack). It is, however, possible to determine the fibre ratio by the use of this method.

5.2.6 X-ray tests

(a) X-ray absorption

This is the most commonly used NDT technique, together with US and dye penetration tests. The X-ray methods used for composites are conventional ones. In the case of composite materials, only X-rays can be used, never γ-rays.

The X-rays used are weak rays obtained with a low voltage source (20–60 kV) and a beryllium window. A fine source is used, $d < 2$ mm, (e.g. a microfocus type X-ray generator). The source is placed at a distance ($H \geq 1$ m) from the surface of the component. On the other side of the composite, opposite the X-ray source a sensitive photographic film is placed or a suitable detector mounted on a bench in such a way that its xy coordinates are accurately specified, as shown in Figure 5.41.

Calibration

For a satisfactory film record either exposure trials must be carried out or a calibration curve used, as shown in Figures 5.42 and 5.43. The latter have been developed by, for instance, CETIM (Centre Technique des Industries Mécaniques, Nantes, France). These figures show some calibration curves for unidirectional composites containing 60% of fibres. These curves depend on specimen thickness, which must be the same for the specimen and calibration material.

Note that carbon and aramid fibre laminates behave in the same way. The X-ray penetration of glass fibre laminates is not so good and a higher-voltage

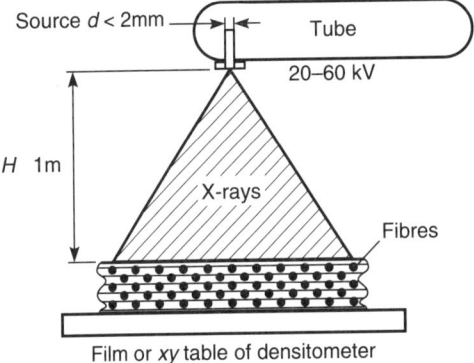

Figure 5.41 Principle of X-ray testing.

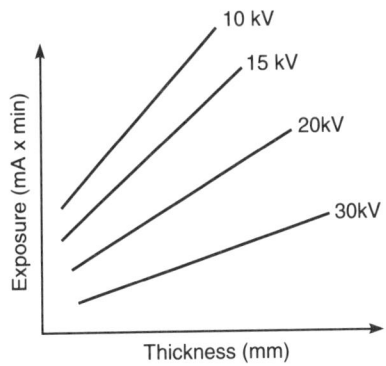

Figure 5.42 Calibration curves for a laminate.

X-ray system is required (20–60 kV); 10–20 kV X-rays are sufficient for carbon or aramid fibre composites.

Owing to their chemical composition (organic carbon chains), carbon and aramid fibres and resins behave in much the same way when exposed to X-rays. It is therefore difficult to differentiate between the two fibres in a laminate using X-rays.

In a unidirectional fibre–resin composite, the attenuation of X-rays is proportional to the weight ratio of fibres up to 60%, as shown in Figure 5.44.

A microdensity analysis of the film can be used, after calibration to estimate the fibre volume ratio.

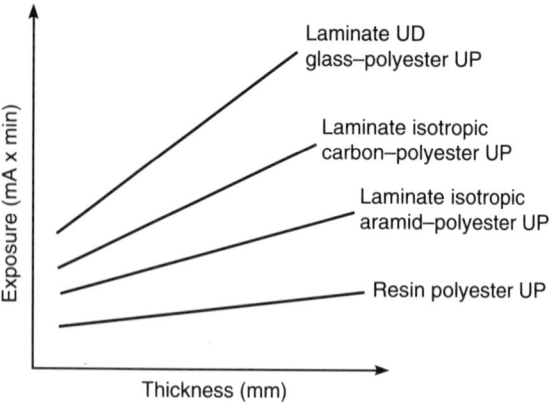

Figure 5.43 Calibration curves for resin and different laminates.

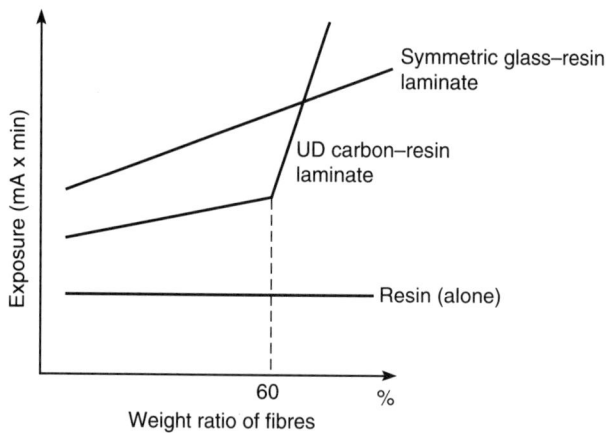

Figure 5.44 Weight ratio of fibres.

Defects that can be detected
X-rays are capable of detecting a number of different defects, among which are:

- inclusions;
- metallic parts within the laminate;
- porosity;
- delamination caused by impact;
- delamination due to machining.

In addition X-rays can be used to measure:

- the thickness of the laminate;
- the alignment and distribution of reinforcing fibres.

It must be pointed out, however, that poor bonding between fibres and matrix cannot be detected.

It is easier to detect porosity in carbon fibre-based laminates since the atomic weights of the main components of the fibres and resin are similar and X-rays are consequently absorbed in a similar way in both components. Testing these materials is therefore akin to working with isotropic materials such as metals.

Sensitivity
It is possible to determine the dimension of porosity-type defects by using a tiered test piece like that shown in Figure 5.45: also known as an image quality indicator (no standard).

This tiered or stepped test piece is made of layers of laminate containing 60% of fibres (by weight). Tiny holes, of increasing diameter, pierce the material. The tiered test piece is exposed to exactly the same conditions as the piece to be tested.

The sensitivity to defect detection is:

$$s = \frac{d}{e},$$

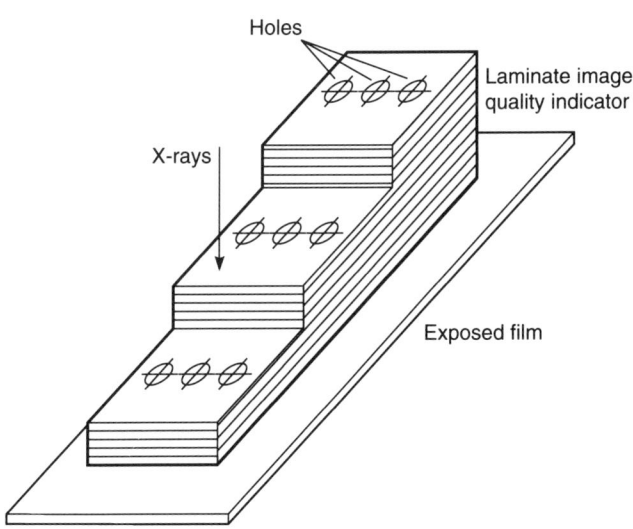

Figure 5.45 Tiered test piece for X-rays.

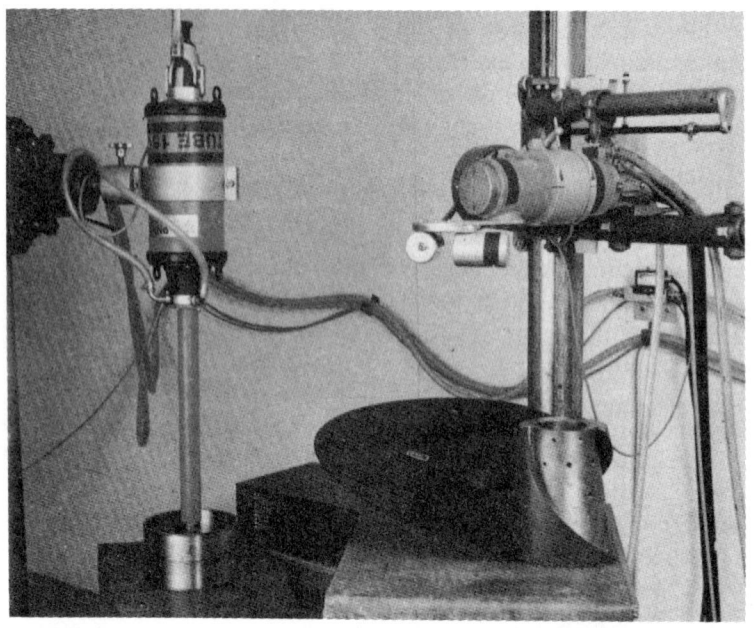

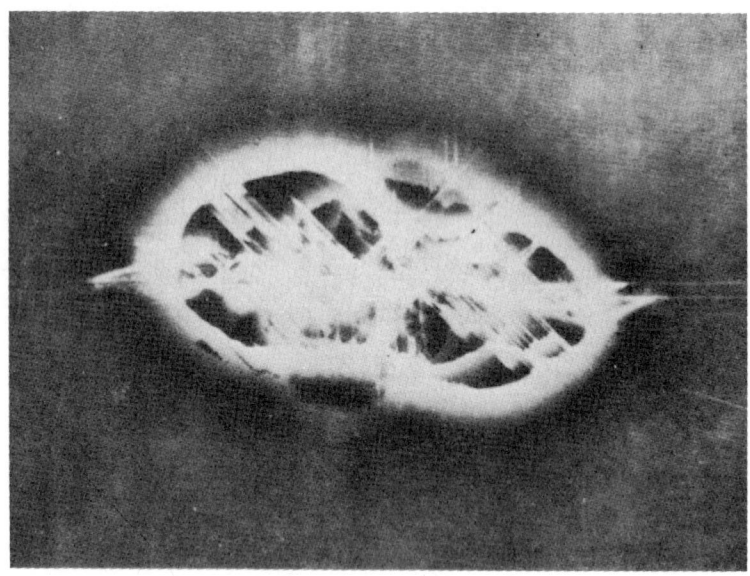

Figure 5.46

d being the diameter of the smallest visible hole (in %) and *e* being the thickness of the laminate tested.

An automatic image processor (Figure 5.46) is used to examine the film and the following can be characterized:

- the morphology and number of defects;
- their distribution in terms of size and diameter.

The principal stages in the processing of the film are:

- filtering;
- setting the limits;
- translation into binary language.

Tangential radiography on curved pieces allows the detection of thin air pockets or delamination. The film is placed behind the test piece, and the X-ray beam is tangent to the component, as shown in Figure 5.47.

Use of radioactive tracers
The use of radioactive tracers or high-density penetrating agents (zinc iodine, barium sulphate, tetrabromoethane, etc.) which get into surface cracks allows the use of X-rays for the estimation of the depth of the crack and the extent of the damage.

(b) Compton scattering
A very stable, narrow (0.5 mm diameter) source of photons impinges on the specimen, as shown in Figure 5.48

Compton scattering involves the elastic collision of a photon with a free electron or an electron which is loosely bound. The energy of the incident photon is $E_0 = h v_0$, h being the Planck constant, and v_0 being the frequency of the incident beam. The scattered photon beam comprises particles of

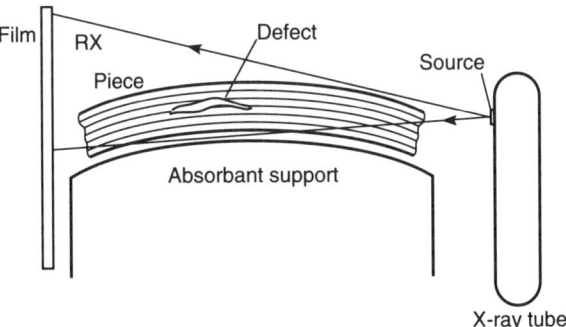

Figure 5.47 Tangential radiography.

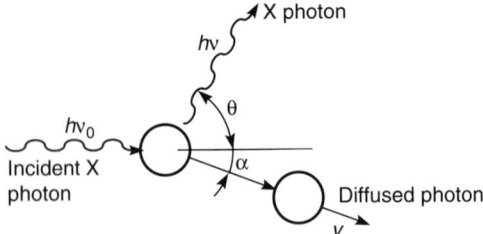

Figure 5.48 Principle of the Compton effect.

energy hv. The incident and final energies of the photon and electron are related by:

$$hv_0 = \tfrac{1}{2}mv^2 + hv.$$

The angle θ through which the beam is scattered lies between $0°$ and $180°$. Retrodiffusion can thus take place. It is used in NDT where retrodiffused photons can be recovered. Their number is proportional to the atomic number and the density of the tested material.

The presence of any defect (delamination, porosity, etc.) in the material significantly modifies the intensity of the scattered X-ray or photon beam compared with the result for sound material, allowing defects to be detected. This method is useful as material can be examined which is not normally accessible (see Figure 5.49).

A detector measures scattering in a direction defined by a collimator. A small volume of material ($0.5 \times 0.5 \times 5\,\text{mm}$) is examined with the source detector assembly moved along the Ox and Oz axis. All defects are

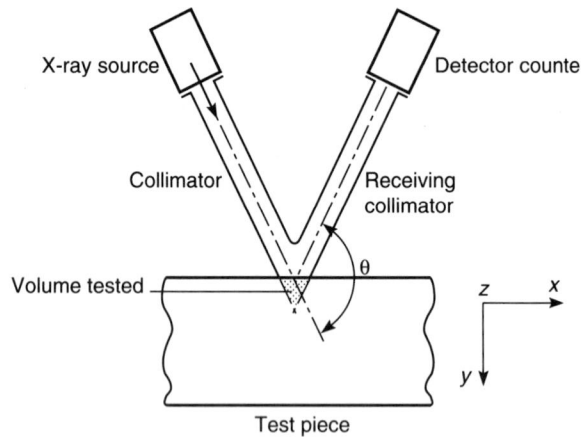

Figure 5.49 The Compton test.

detected even though they may lie in a plane. Maximum scattering is related to the density and average atomic number of the test piece. The technique has been used, for example, for examining rocket motor cases (USA).

It is possible to determine density remotely. This is especially useful when the test piece is massive or when it is a sandwich structure.

This method can be used to detect porosity and determine the fibre content (if there is no porosity).

5.2.7 Neutron radiography

(a) The limitations of X-rays

The absorption of hard electromagnetic radiation (X- rays, γ-rays etc.) follows the well-known law:

$$I = I_0 e^{-\mu/\rho} \times \rho x,$$

I_0 being the incident intensity, I the transmitted intensity, μ/ρ the absorption or attenuation coefficient of the radiation, ρ the density of the material, and x the thickness of the material through which the radiation passes, as shown in Figure 5.50.

The coefficient μ/ρ varies with the energy of the incident photons, i.e. their wavelength λ according to the following approximate relation:

$$\mu = kZ^3\lambda^3\rho,$$

Z being the atomic number of the material through which the radiation passes, and k is a coefficient.

For an initial approximation, we can write $Z = A/2$, where A is the atomic mass of the element.

The curves $\mu/\rho = F(Z)$ or $F(A)$ show abrupt decreases for certain values of λ and the coefficient k is only constant between two successive discontinuities, as shown in Figure 5.51.

The approximate value of the absorption coefficient μ/ρ increases with the atomic mass A, the density ρ and the wavelength λ, or the energy, of the

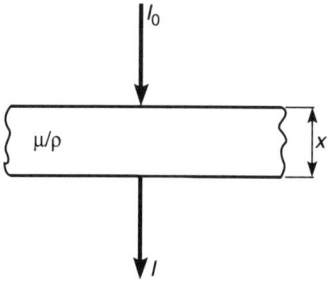

Figure 5.50

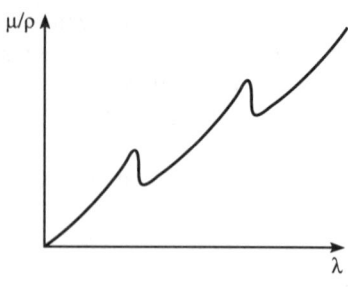

Figure 5.51

radiation. Thus it appears that light materials, with low atomic numbers are essentially transparent to electromagnetic rays. This is the case for composite materials; the resin is based on carbon: $A = 12$, $Z = 6$, $\rho = 1.8\,kg/dm^{-3}$, or hydrogen with $A = 1$, $Z = 1$.

On average, the density of the resin is $\rho = 1.2\,kg/dm^{-3}$. Fibres are based on carbon or glass (silica), SiO_2. For Si: $A = 28$, $Z = 14$, and for O: $A = 16$, $Z = 8$. The density of silica is $2.2\,kg/dm^{-3}$. Alternatively the fibre may be based on boron $A = 11$, $Z = 5$, $\rho = 2.4\,kg/dm^{-3}$.

Since the values of the various densities, atomic numbers and atomic masses are very close, the variation of μ/ρ is minimal for the X-rays which are unable to distinguish resin from fibre or voids. Thus conventional X-ray methods are limited in checking for these defects.

(b) The advantages of thermal neutrons

Examination using thermal neutrons is a useful alternative to using X-rays for low-density materials containing substantial amounts of carbon, oxygen and hydrogen atoms. Using neutrons, it is possible to distinguish between boron and carbon. The neutron absorption coefficient for boron is 300 times that for carbon. Neutron radiography enables boron fibres in a resin matrix to be detected and their orientation and fibre volume loading to be determined.

(c) The disadvantages of thermal neutrons

Unfortunately, the technique requires a specially adapted nuclear reactor to supply the neutron flux. In addition, only small composite pieces can be tested. However, the development of portable neutron sources is under consideration. The successful development of one of these would give a great impetus to the technique.

5.2.8 IR thermography

This method is very interesting because it is a large-scale and rapid technique.

(a) Principle

All materials radiate energy in the form of electromagnetic radiation over a wide wavelength spectrum. The energy radiated per unit area, P, is given by the Stefan–Boltzmann law, namely:

$$P = \varepsilon \sigma T^4,$$

where ε is the emissivity of the surface, σ is the Stefan–Boltzmann constant = $5.67 \times 10^{-8}\,\mathrm{Wm^{-2}k^{-1}}$, and T is the temperature in degrees Kelvin. The emissivity, ε, is a fundamental parameter in IR thermography; it is an intrinsic characteristic of the surface of the material. By definition ε always lies between 0 and 1. Some common values are:

- aluminium, $\varepsilon = 0.05$;
- steel, $\varepsilon = 0.1$;
- glass fibre resin composite, $\varepsilon = 0.9$.

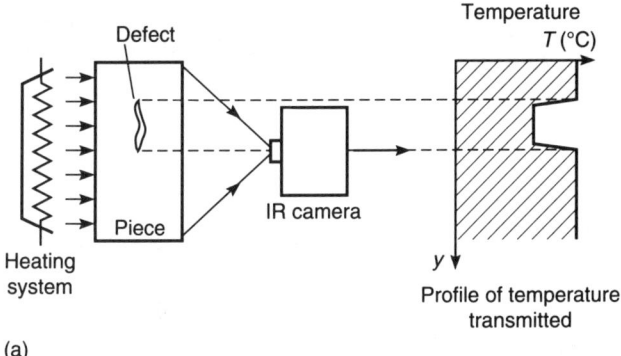

(a)

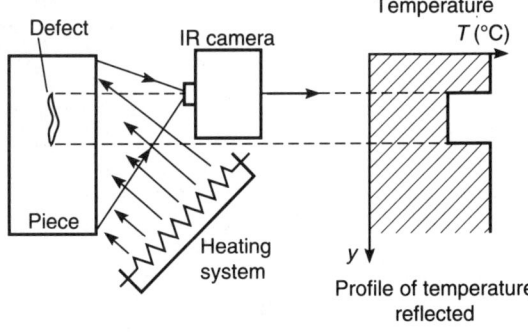

(b)

Figure 5.52 IR thermographic examination: **(a)** direct heating; **(b)** indirect heating.

 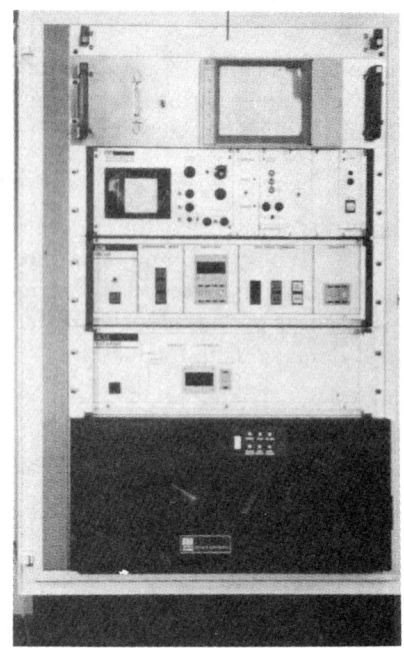

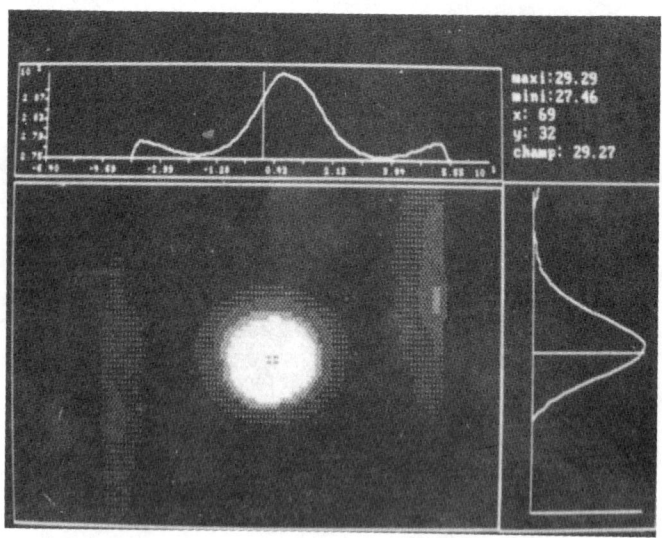

Figure 5.53

Composite materials which possess a high emissivity can be tested easily by IR thermography. If the emissivity of the material is too low, a coating of material possessing a high emissivity must be deposited on the surface of the test piece.

The following equipment is required for an IR thermographic examination:

- an IR camera as a detector;
- a mechanical scanning system;
- a digital signal processing unit.

The camera views the IR beam through suitable lenses and prisms. The beam is filtered and focused on a detector (indium antimonium) placed in contact with a cryogenic container filled with liquid nitrogen ($-196\,°C$). The detector produces an electronic signal proportional to the power, P, radiated by the object. This signal is amplified and processed digitally, as illustrated in Figures 5.52 and 5.53.

Two techniques are used: either the test piece is heated directly from the opposite side to the detector, or the test piece is heated from the front.

(b) IR absorption thermography

Two-stage technique
First, the test piece is heated evenly using IR lamps placed in front of it as shown in Figure 5.54. It should be noted that it is difficult to obtain even heating over the whole surface. Secondly, the heating is switched off and the test piece allowed to radiate. Any defect or anomalous area which does not possess the same thermal characteristics as the undamaged part of the test piece radiates in a different manner which is detected by the IR camera. Thus

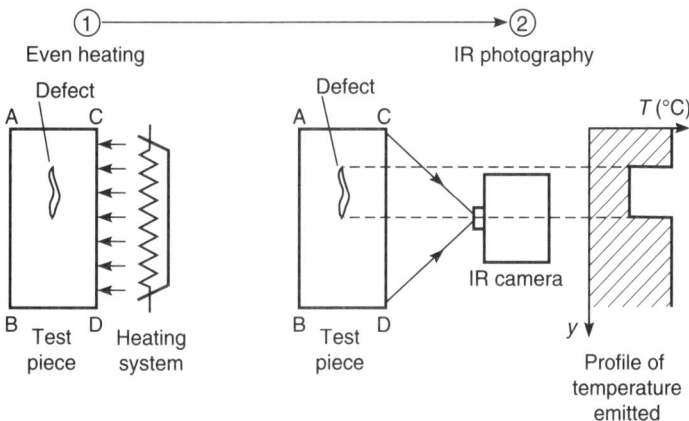

Figure 5.54 Two-stage technique.

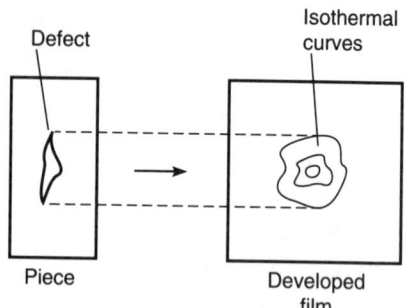

Figure 5.55

variations in the radiation pattern of damaged and sound material are detected.

To summarize: this is a two-stage method, illustrated in Figure 5.54, of testing a possibly large specimen involving heating time and a period during which radiation from the test piece is measured.

To assess the radiation pattern either a large-scale picture of the specimen surface is obtained, as in Figure 5.55, or the surface is systematically scanned by the IR heating lamps/camera assembly in order to study the whole surface. Pictures of different areas can be combined to give an overall impression of the artefact.

It may be preferable while still using the IR absorption method, to run the IR tubes/camera assembly in front of the test piece.

Description of the technique
The IR heating lamps are moved slowly and at constant speed in front of the composite test piece and the camera follows at the same speed. The distance between the two remains constant. Image processing electronic equipment processes the pictures line by line and these are recorded on a video-cassette recorder (VCR), which allows the recorded data to be studied later and the results of the tests to be filed; see Figure 5.56.

Thermograms show defects which interfere with the propagation of heat through the sample, for instance air bubbles or voids, cracks, delamination and debonding of sandwich skins.

Note that when sandwich structures are tested, it is impossible to distinguish between delamination within the skin and inadequately bonded skins.

Until now, with monolithic composites, it has not been possible to assess the depth of the defect using this technique and a complementary ultrasonic test has had to be carried out on the area containing the defect.

Research carried out in France at the Compiegne Institute of Technology indicates that automatic digital image processing may now allow the volume of the defect to be assessed.

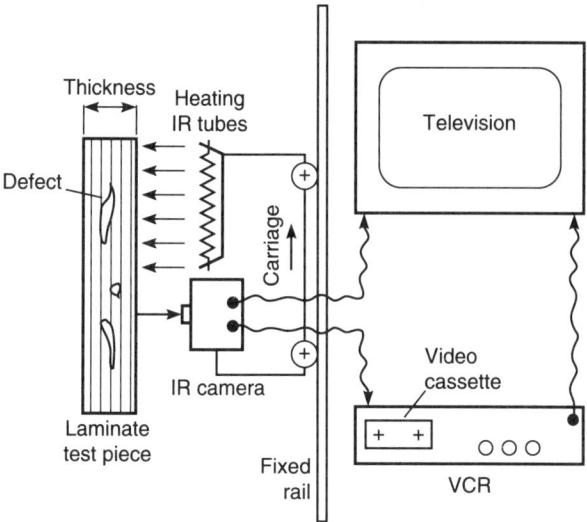

Figure 5.56 Method consisting of running the heating system in front of the test piece and monitoring absorption patterns.

Note that the use of temperature-sensitive liquid crystals spread on the surface of the laminate facilitates the testing, as these change colour from red to blue on heating.

(c) IR transmission thermography

The test piece is evenly heated either by IR lamps in front of it or heated sequentially by scanning with IR lamps. This is a one-stage technique and is illustrated in Figure 5.57. The heat is conducted through the test piece and the flux detected at the opposite face using an IR camera or a suitable TV camera.

(d) IR vibrothermography

The piece to be tested is subjected to high-frequency mechanical vibrations using a piezoelectric vibrator. Internal friction and deformation in the composite causes thermal energy to be radiated. The damaged parts radiate a great deal more energy than the undamaged areas and this difference is detected with an IR camera. This technique is becoming increasingly popular.

5.2.9 Laser holographic interferometry

This method is well adapted to the study of large composite CNDs. The French aeronautical industry uses this method to test all the composite helicopter blades it produces (these are about 10 m in length) at Eurocopter Co.

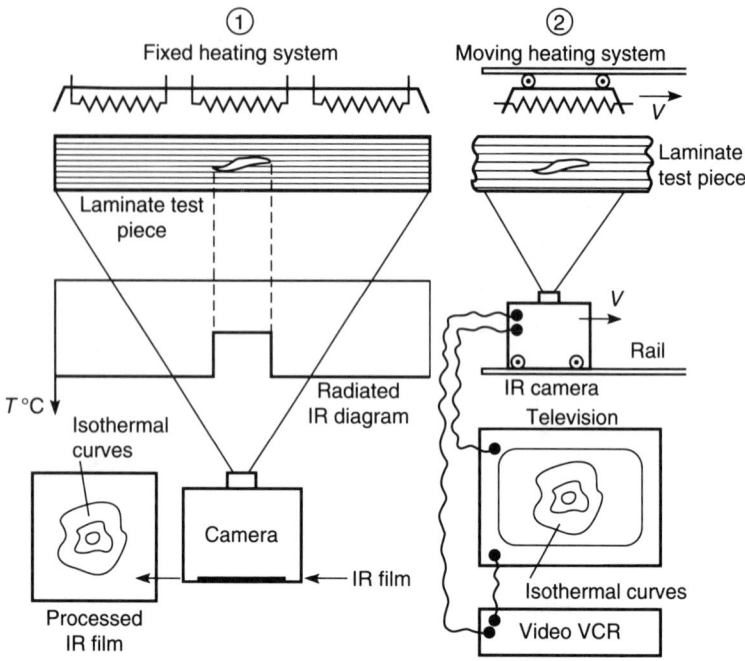

Figure 5.57 Transmission IR thermography: one-stage technique.

(a) Principle

Two pictures, one of a balanced, unloaded, component and the other of the component loaded, are superimposed. In the latter case the specimen is slightly deformed. The light from the two pictures interferes and generates a three-dimensional interference pattern. The distance between the fringes is $\lambda/2$ (with argon $\lambda = 5145\,\text{Å}$).

Any defect within the piece produces a local anomaly affecting the overall deformation of the test piece, the fringe pattern is disrupted in the vicinity of the defect and the latter detected as in Figures 5.58 and 5.59. Collapse measuring about 10 fringes can be detected (3 μm) on defect areas of 10 mm diameter.

(b) Applying load on the specimen

Jacks can be used and a reduced pressure of 7–10 mbars can be used either by partially evacuating the test room or, if this cannot be sealed, by using a vacuum bag. Deformation can be produced by uneven heating.

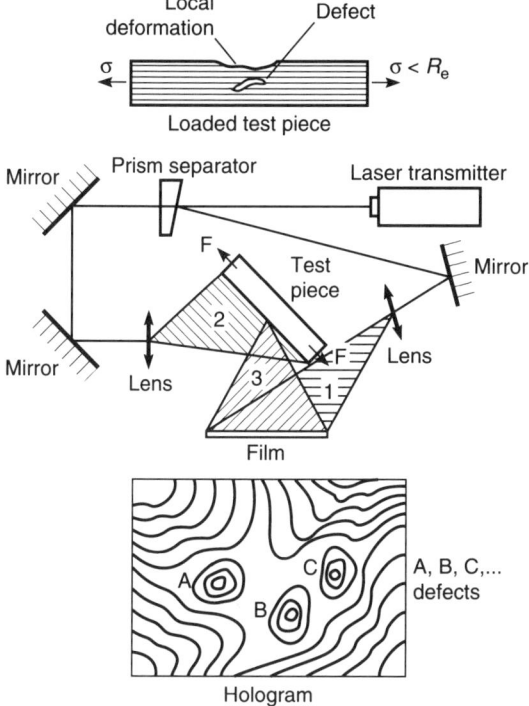

Figure 5.58 Principle of holography.

(c) Results

The laminate is placed under load to cause deformation. The deformation of the surface is more acute in an area containing defects. The local anomalies affecting deformation patterns are seen in three-dimensional holograms when the interference spectra of laser beams 1 and 3 are analysed.

The results can be analysed both qualitatively and quantitatively. The qualitative and visual analysis of the fringe pattern allows us to determine that the defects (A and B in the figure) are near the surface of the laminate (1 to 3 mm deep). The different types of defects can be classified and associated with characteristic interference patterns. Quantitative analysis is now possible using data processing equipment which determines the deformation area.

(d) Future of the method

This technique is now commonly used for testing sandwich structures, structures such as rocket engines, and helicopter blades. A vibration-free laboratory darkroom is required. The test piece is placed on suction pads on a slab of granite mounted on pneumatic jacks and subjected to the beam

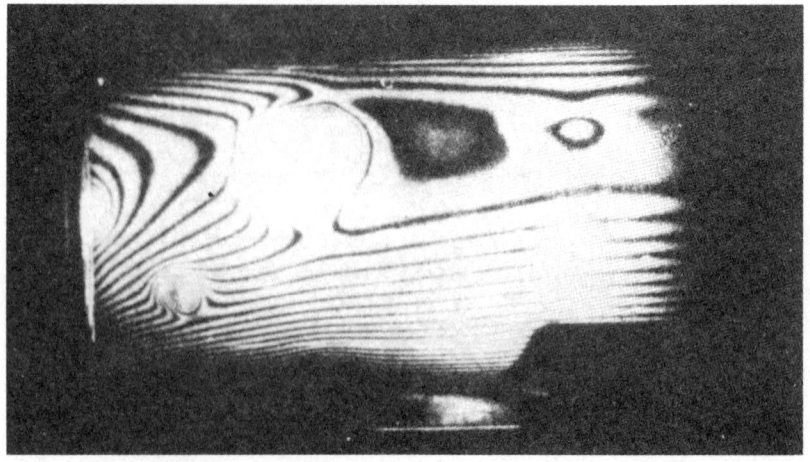

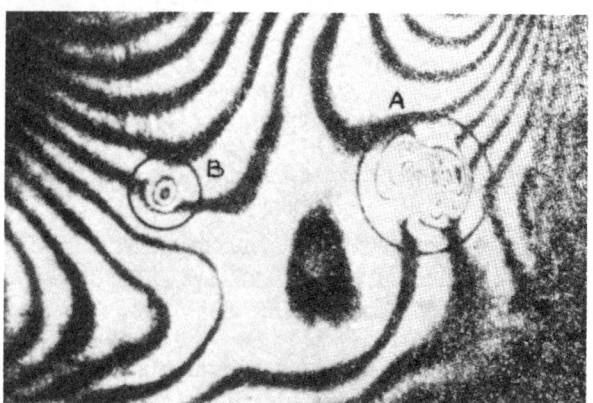

Figure 5.59

of a continuous-flow gas laser (argon). The ancilliary equipment needed for the test is complex and expensive and small firms cannot afford it. Eventually high-energy pulse lasers, which can be used in the workshop, will be employed. Holograms will then be produced more quickly and in daylight.

5.2.10 Acoustic testing (Standard AFNOR A09360)

Any composite material subjected to a stress emits acoustic signals, which may or may not be audible, when it is deformed locally or when microcracks are produced (due to the rupture of the matrix, or reinforcing fibres, or delamination) or when a laminate skin debonds from the honeycomb or foam core on a sandwich structure.

Listening to this noise in real time using piezoelectric sensors makes it possible to localize the source, and to give an interpretation of the events taking place in the material. Numerous factors affect acoustic emission from composites and interpreting the signals is not easy. The analysis is much more difficult for a composite than is the case for an isotropic material.

Acoustic methods are used in non-destructive testing though the specimen is in fact damaged. A load is applied up to level σ_1, creating a signal or

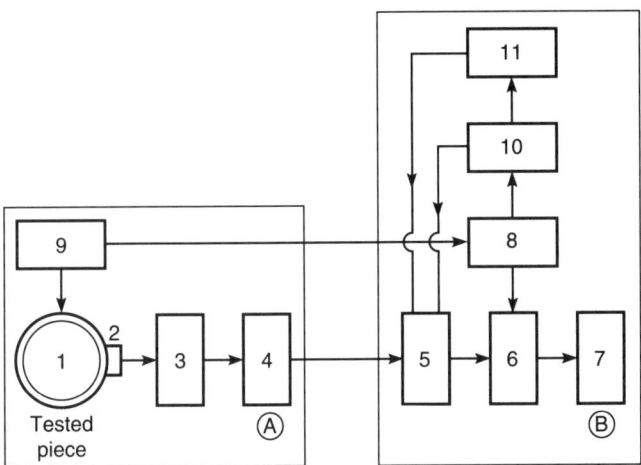

(A) Signal acquisition zone

(B) Signal processing zone

(1) Test piece

(2) Noise sensor

(3) Background noise filter

(4) Amplifier (100 dB)

(5) Separator

(6) Counter

(7) xy - drawer

(8) Clock

(9) Loading system

(10) Oscilloscope (visualization)

(11) Output for loudspeaker or heterodynage

Figure 5.60 Instrumentation for the acoustic test.

noise. The load is then removed and reapplied at a level $\sigma_2 > \sigma_1$. This causes a different signal to be produced. A second experiment can be carried out after the component has been used for some time in order to determine:

- whether further damage has occurred;
- whether new breaks or fractures have been produced and to determine what the defects are.

The instrumentation needed is shown in Figure 5.60.

A given, local, deformation ε in time t corresponds to a stress σ and a signal whose amplitude A and frequency spectrum are shown in Figure 5.61.

Signal processing is difficult and complex because the phenomena observed are random. Consequently, counting is the method most frequently used. A discriminator is employed in order to screen out non-significant background noise, only the signals whose amplitude is greater than a given value AD are observed. The principal frequencies between 50 and 300 kHz indicate: the break of resin for 30 dB; the lamination for 40 dB and the cracks of fibres for 80 dB (see Figure 5.62).

The matrix cracks about $0.2 R_m$ of composite. With three sensors C_1, C_2, C_3 carefully placed along the loaded composite component, it is possible to measure the time taken by the signals to travel from the source (defect) to the different sensors. Automatic triangulation procedure enables the position of the defect that has caused the signal to be determined. This is a difficult task because of the high signal attenuation in composites. This method is a

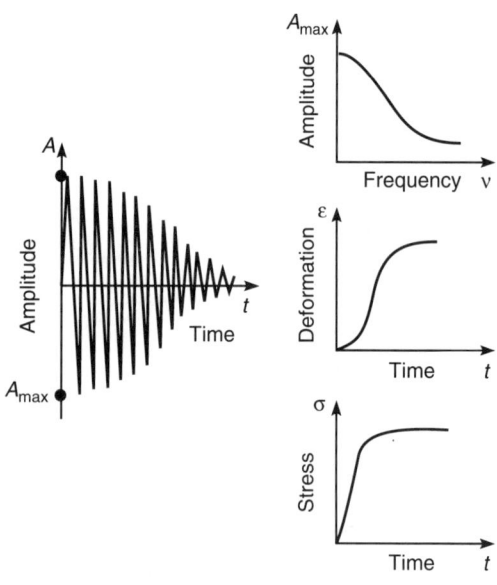

Emitted signals

Figure 5.61

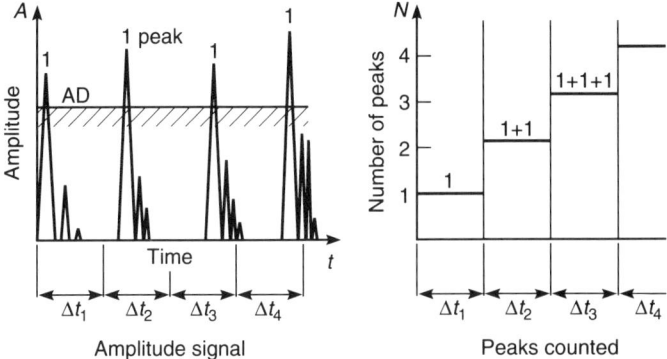

Figure 5.62 Signal processing.

large-scale one which should be followed up by local analysis (X-rays, ultrasonics or thermography) for further details. It is used for the certification of glass or aramid fibre composite pressure vessels.

It would appear that the method has only a modest potential for composite materials. Its large-scale use should be limited to the study of signals emitted by the composite when the latter is subjected to different load levels.

5.2.11 X-ray scanning tomography

(a) Limits of the method

The X-ray scanning system developed for medical purposes has considerable promise for testing composite materials.

It is currently under evaluation for the detection of:

- voids and inclusions with a minimum size of 0.05 mm^2;
- delamination;
- cracks with a minimum width of 0.1 mm.

The industrial use of this technique may prove more difficult than its medical use.

Depending on the material to be tested, much more powerful X-ray sources may be needed – up to 400 kV. As composites are not so absorbent, the X-ray source voltage could probably be reduced to 160 kV. The investment needed is large and beyond the capacity of small companies.

The method has been tested with prototype equipment developed in France at CEA and Intercontrole Society has used it for industrial applications.

(b) Principle of X-ray scanning tomography

A very narrow X-ray beam (1 mm^2) of intensity I_0 traverses a segment AB of the material to be examined, as shown in Figure 5.63. The beam is

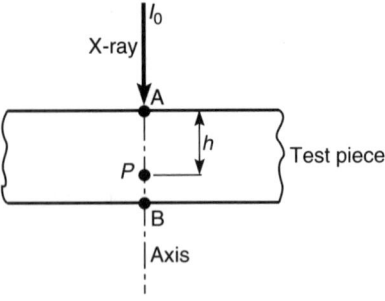

Figure 5.63

partially absorbed, the attenuation being determined by the thickness h of the material and by the absorption coefficient μ. The coefficient $\mu \, (\text{cm}^{-1})$ depends on the nature of the material and varies with the energy of the X-rays used. The ratio $\tau = \mu / \rho$ where ρ is the density of the material is often employed. For 400 kV X-rays $\tau = 0.99$ for a resin. Assume that the transparency τ varies from one area to another. When the beam goes through a given section, $I = I_0 \exp(-\mu h)$. For material which is not homogeneous ρ is a function of h:

$$ I = I_0 \exp \left[-\int_A^B \frac{\mu}{\rho}(\rho(h)) - \mathrm{d}h \right], $$

where h refers to a point on the path of X-rays and $\rho(h)$ is the density of the material at point P. (Each point on this path has a value of τ which is a function of h; see Figure 5.63.)

Assume that a monolithic laminate has an average absorption coefficient μ. In a region where there is a defect such as a void or air bubble or delamination, the absorption coefficient will be effectively zero.

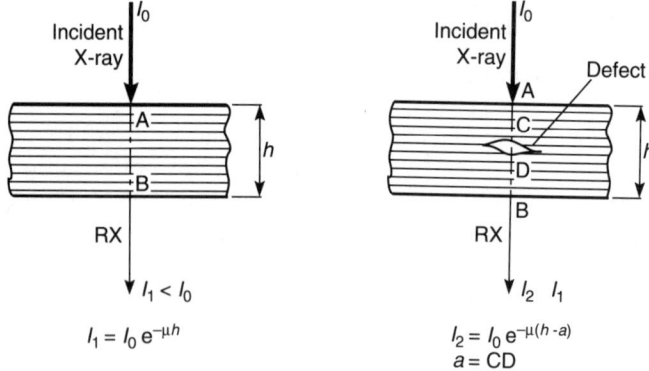

Figure 5.64

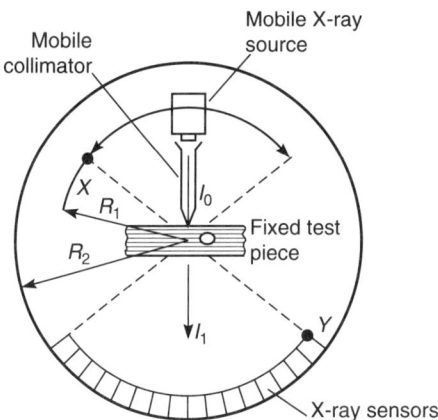

Figure 5.65 Tomography principle.

The intensity of the transmitted beam of X-rays is I_2 and the contrast between an undamaged area and an area containing defects is I_2/I_1. The attenuation A is equal to $I_0 - I_2$ in the area containing the defect and $I_0 - I_1$ in the undamaged area, as in Figure 5.64.

The scanner measures this attenuation by monitoring the transmitted signal (I_1 or I_2) using a sensor 100 times as sensitive as X-ray film. The resolution is <0.1 mm.

(c) The industrial scanner

An industrial scanner differs technically from a medical one. The equipment was developed in France by Intercontrole Society. It was noted in the preceding section that the intensity I_2 or I_1 of the transmitted beam is the summation of all the contributions from small sections along the path of the beam. Unfortunately, the value of the attenuation μ at point P in the test piece with local density ρ cannot be measured easily.

Tomography: principle
It is possible to determine the value of μ by using the X-ray beam or the test piece in three operations to analyse the properties of a small section of the material of height h.

First stage: translatory movement of the source-sensor assembly along axis OX perpendicular to the X-ray beam in order to scan the whole object to be tested.

Second stage: rotation of the source-sensor assembly around the OZ-axis through a small angle $\Delta\alpha$.

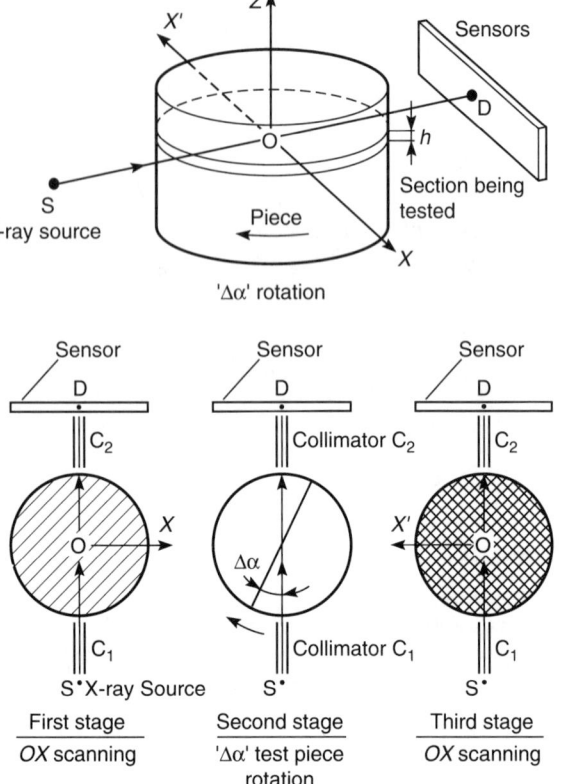

Figure 5.66 Tomography principle.

Third stage: opposite translatory movement along axis OX' of the source-sensor assembly.

These operations are carried out successively to perform a total 360° rotation.

The combination of translations and rotations creates a fine spatial picture of the object in the vicinity of the material at h, as shown in Figure 5.66.

Test bench
In practice, the piece to be tested is placed on the centre of a plate, shown in Figures 5.67 and 5.68, which is turned slowly through an angle $\Delta\alpha$ and which is capable of translatory movements along the OX and OX' axes. The X-ray source S, the scintillation sensors D, and the input and output collimators C_1 and C_2 are all lined up along the vertical, fixed, supports of

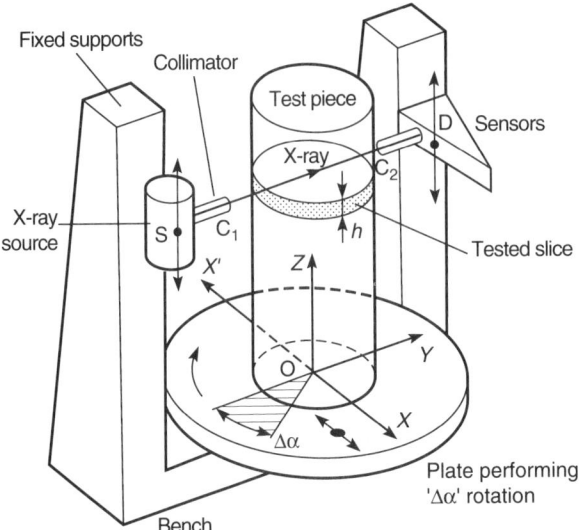

Figure 5.67 Intercontrol® TAO bench.

the machine. The sensors are scintillation counters coupled to photomulti-pliers.

The collimators C_1 and C_2 reduce the X-ray beam to a minimum diameter of 0.2 mm (maximum 15 mm). This makes it possible to determine the thickness h of the section to be tested. After scanning the section, the source sensor assembly, and collimators C_1 and C_2, are gradually moved upwards along the OZ-axis and in this way the entire specimen can be scanned, section by section.

Note that the plate can be moved along an axis OY so that the test piece can be brought closer to the X-ray source or to the sensors D.

Data processing
The X-rays which have traversed the specimen are incident on the scintillation counters D. This signal is amplified, digitized and processed by computer using the appropriate mathematical software. Within a few minutes (30 minutes for large components) the computer can process X-ray attenuation data for all the sections of the specimen.

An imaging system makes it possible to:

• give a graphic plot of X-ray attenuation or the intensity of the transmitted signal;

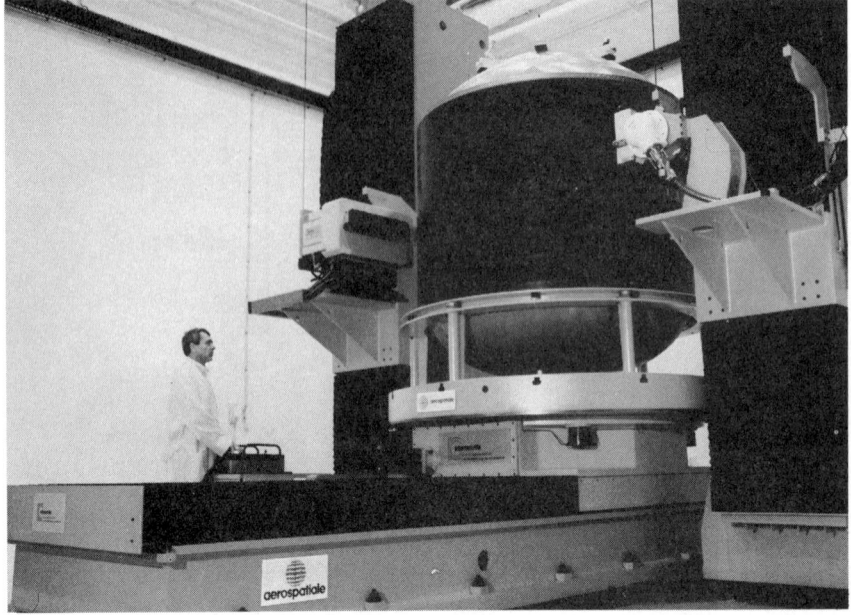

Figure 5.68 Intercontrôle® TAO bench.

- reconstruct the image of the section and display it on TV monitors;
- store the results and images on tape or discs for future reference or use.

(d) Results

The intensity of the transmitted X-ray beam is indicated by different shades of grey or different colours and displayed on a TV monitor. The different shades of grey are expressed in terms of Hounsfield (H) units. These range from -1000 for air to $+1000$ for bones, zero corresponding to water. Variations in H can be measured down to 3 units, indicating the sensitivity of the method. The narrower the range of intensity studied, the higher the discrimination in H. Using the image, variation in H can be studied at one point or one area and the results displayed in histogram form. Dimensions can also be measured.

Figure 5.69 shows an intensity pattern obtained from a delaminated area and from an undamaged area of carbon fibre epoxy composite. In the delaminated area, H is weaker than in the defect-free area. Moreover, the histogram of results for the undamaged area is much sharper than that obtained for the damaged area. The results can be stored on a disc or tape as shown in Figure 5.70.

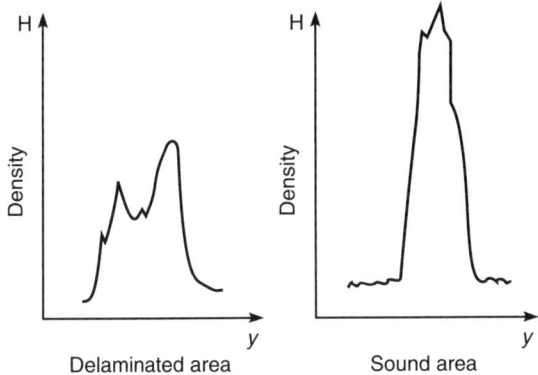

Figure 5.69

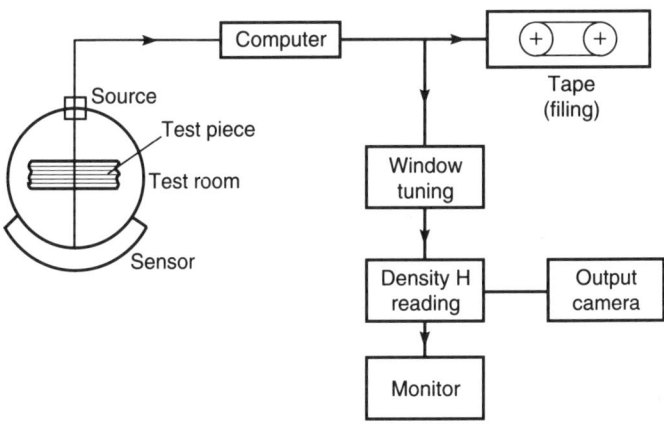

Figure 5.70 Scanning equipment.

(e) A device with a future: the imatron

The imatron is a video-scanner. It does not involve any moving parts with
the exception of the test piece. A very large electron gun beams electrons,
focused by magnetic coils, on to tungsten (W) target rings placed around the
test specimen. These target rings are excited by the incident electrons and
emit X-rays. The X-ray beam is then focused on to two rings of sensors
inside the rings, as shown in Figure 5.71. Each ring is scanned rapidly by
electronic focusing and 10–30 images per second are obtained.

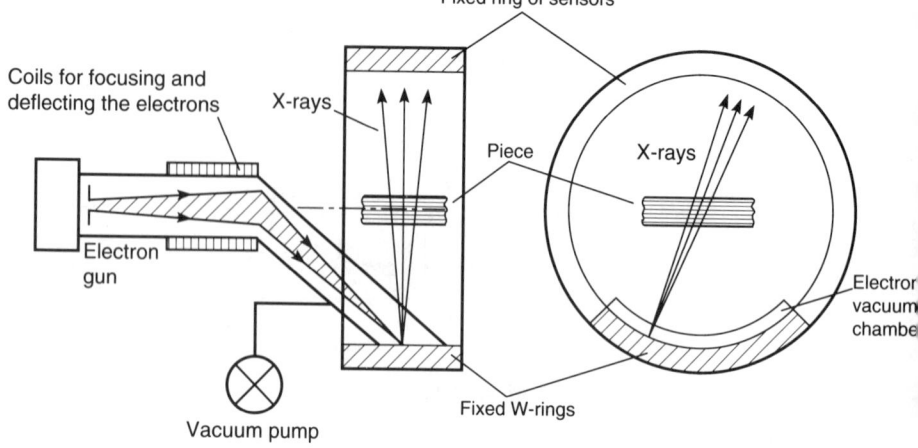

Figure 5.71 The Imatron.

5.2.12 Other non-destructive methods

(a) Methods using stimulated stress waves

In about 1977, NASA developed a method which combines acoustic emission and ultrasonic testing. An ultrasonic source sends a set of signals through the composite at regular intervals. If g is the time during which ultrasonic vibrations are counted, r is the recurring period of the vibration spectrum and N is the number of oscillations or peaks over the set threshold, then the method is as illustrated in Figure 5.72, and the following equation holds:

$$S = grN \quad \text{(wave factor)}.$$

This can apparently be linked to the break resistance of the composite.

The equipment which was developed by NASA allows the user to determine the probable failure zone in a composite. Failure is said to occur when the stimulated wave factor S is at its lowest, as shown in Figure 5.73.

(b) Proof pressure testing of tanks

The pressure in the tank is raised to the proof pressure P_Q, which is higher than the standard service pressure P_S, but lower than the theoretical pressure P_E at which the structure would reach the limit of its elastic response σ_e. The proof pressure P_Q may be used both for a proof and for quality control testing.

$$P_Q > P_S \quad \text{and} \quad P_Q < P_E.$$

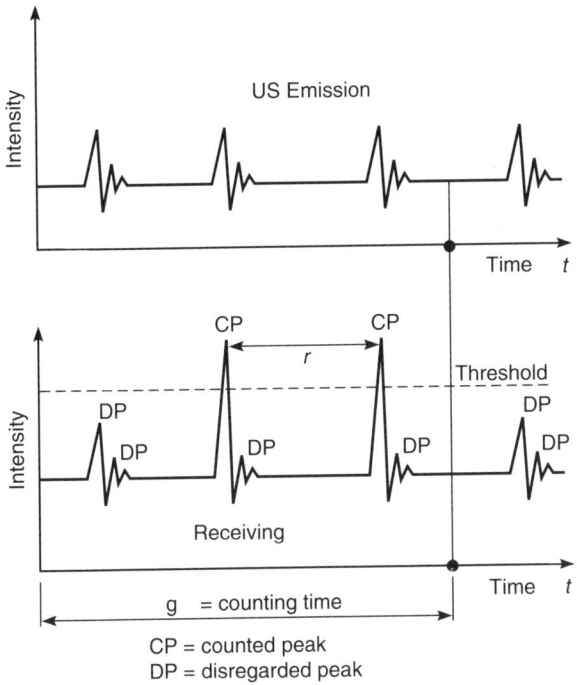

Figure 5.72 Counting the peaks in stimulated stress waves.

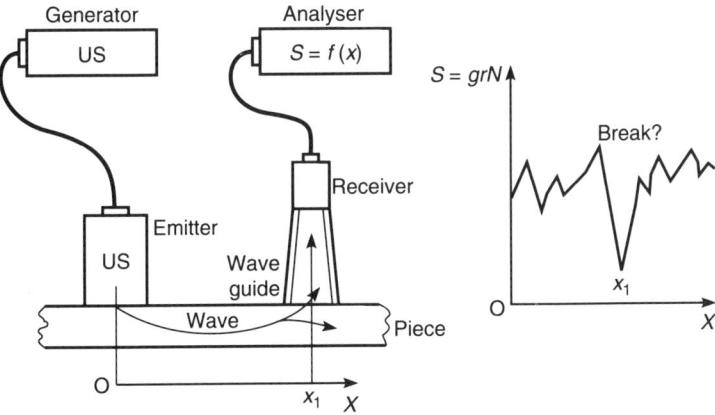

Figure 5.73 Determining the break zone in stimulated stress waves.

The method may be complemented by using acoustic emission to locate the damaged areas.

(c) Potentiometric tests

The electric resistance of a conducting composite (such as carbon) is measured in the thickness of the composite as shown in Figure 5.74. The laminate is placed between two electrodes, and the current and voltage measured. The conductivity may be unduly low if laminate layers are in close contact (due to misaligned fibres for instance).

(d) Testing using eddy currents

The use of high-frequency currents in inductive coils running along the surface of carbon fibre based conductive laminates generates induced currents known as eddy currents in the carbon fibres. Eddy currents, owing to their high

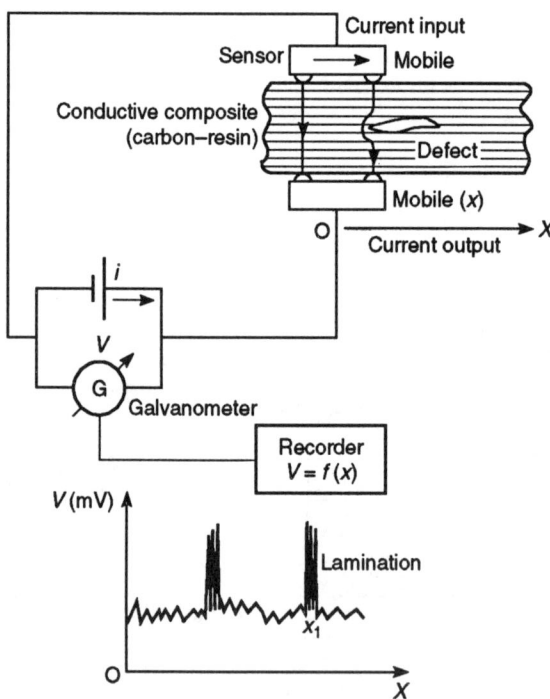

Figure 5.74 Potentiometer control.

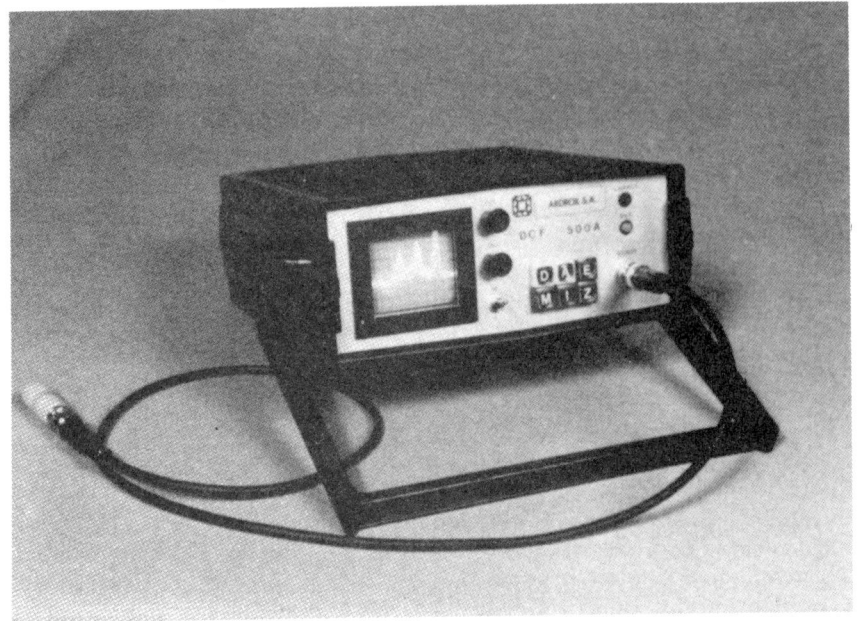

Figure 5.75

frequency, do not penetrate deeply into the structure and therefore only affect a few layers of the composite.

Any rupture, delamination or alteration of the reinforcement in the surface regions thus becomes apparent. Testing can be carried out automatically. This kind of test procedure is exclusively qualitative.

Composite materials were initially assumed to be homogeneous, but this is not the case. Only carbon fibres are conductive so that the laminate can be regarded as made up of capacitator units (insulating material between fibres) and resistance units (fibres). The heterogeneity of a composite material accounts for the behaviour of carbon fibre composites when eddy currents are induced in the composite.

The device, illustrated in Figure 5.75, is based on measuring the impedance and capacity variations of the coil which is connected to an oscillator. Any variation in the impedance or capacity corresponds to a variation in the amplitude and frequency of the signal which is produced. A detection system displays variations on a small screen linked to a memory unit.

(e) Tapping test (no standard)

It is well known that a flaw-free piece of metal produces a high, clear sound when hit. This old technique is still used to check the wheels of railway carriages for possible fatigue flaws. The performance of the inspector is tested from time to time with an audiogram.

The method requires a certain degree of experience. It is, however, reliable enough to be used to check composite materials. The technique is known as 'tapping' and is carried out with a special small hammer in a second sound-proofed room.

In France all helicopter blades made of composite materials by Eurocopter are first holographed and then tapped with a small hammer in order to localize any defective areas (diameter $\phi \geqslant 7$ mm) such as:

- delamination in a monolithic laminate or in the skin of a sandwich structure;
- defective adhesion between the skin and the core of a sandwich structure.

The method is simple, fast and cheap.

The composite piece is mechanically insulated from the ground and is tapped lightly in order to excite mechanical vibrations that may then be analysed. The analysis of the vibration is carried out by:

- listening to the sound produced;
- monitoring the sound automatically with microphones or direct-contact piezoelectric sensors. The necessary devices are available on the market.

The equipment consists of a microphone, a sensor for data acquisition, a signal processing device, and software for analysing the results.

The sound produced by the piece is analysed either over a given period of time or across the frequency spectrum produced.

Correlating the results obtained with types of defect is not easy, and further studies are currently being conducted. They are focused on the first mode of vibration. It has been noted that the frequency of this mode of vibration increases with the fibre fraction, and decreases with porosity or void content. One can also use the ratio between the amplitude of a mode of vibration, rated as weak, and the amplitude of a mode of vibrations rated as high but having the same parity. This ratio would increase with the porosity ratio.

(f) Microwave testing

The use of microwaves has led to a new family of NDT techniques now being developed.

Aerospatiale Society, France carries out a continuous evaluation of the resin content of aramid fibre composites by measuring the dielectric properties with microwaves. The distribution of fillers within a composite matrix can also be evaluated. At present there is no specific scientific relation between dielectric properties and the size or proportions of defects. To use the technique for this purpose, comparison standards must be employed.

Table 5.2 (a)

	Too much resin	Folds	Porosity
X-ray	Detectable	Detectable	Detectable
Dye penetration test	Not detectable	Not detectable	May work if porosity is near the surface
Ultrasonics	Not detectable	May be detected	Probable detection
Acoustic emission	Unsuitable method; may work, however	Unsuitable method; may work, however	Not detectable
Thermography	Not detectable	Not detectable	Not detectable but may affect the picture
Holography	Not detectable	Not detectable	Detection is possible in some materials
Shock response spectral analysis	?	?	?

Table 5.2 (b)

	Inclusion	Delamination	Crack
X-ray	Detectable	Detectable	Possible
Dye penetration test	Not detectable	Detectable if delamination affects the surface	Detectable if crack intersects the surface
Ultrasonics	Possible	Detectable	Detectable
Acoustic emission	Unsuitable method, but may work in some cases	Detection becomes possible if delamination occurs during testing	Detection becomes possible if cracking occurs during testing
Thermography	Very improbable that detection is possible except in some special cases	Detection possible in some cases	Very improbable that detection will occur
Holography	Very improbable that detection is possible except in some specific cases	Detectable if delamination causes deformation during test	Detectable if crack causes deformation during testing
Shock response spectral analysis	?	Probable detection, currently being studied	Probable detection, currently being studied

Table 5.2 (c)

	Lack of adhesive between skin and core	Inadequate adhesion	Inadequate cohesion
X-ray	Detectable	Not detectable	Not detectable
Dye penetration test	Detection possible if joint line reaches the surface – this method is generally unsuitable	Not detectable	Not detectable
Ultrasonics	Detectable	?	?
Acoustic emission	Detection very improbable but may be achieved in some cases	Detection very improbable but may be achieved in some cases	Detection possible
Thermography	Detectable	Not detectable	Not detectable
Holography	Indirect detection is possible if the absence of adhesive affects the surface	Indirect detection is possible if in-adequte adhesion affects the surface	Detection impossible
Shock response spectral analysis	Detection may be envisaged	Detection very improbable	Detection improbable

5.2.13 How to look for specific defects

Table 5.2 allows the selection of the most appropriate NDT techniques for detecting different types of defects.

5.3 GEOMETRICAL CONTROL OF THE FINISHED PRODUCT

5.3.1 Purpose of geometrical control

The purpose is to ensure that the dimensional specifications detailed in the contract between the manufacturer and the customer have been met. This contract contains the schedule of conditions and the plans. In essence the process involves conventional metrology applied to components made of composite materials.

5.3.2 Main geometrical tests

These are as follows:

- Thickness of the laminate.
- Shape of the surface.

- Relative positions and orientations of surface areas (assuming that the composite is not a plane sheet).
- Warping of the surface.
- Surface appearance and roughness.

All the above have to be evaluated to check that the product meets specifications and can, if required, be assembled properly to produce the final component.

5.3.3 Thickness of the component

(a) Test conducted on the edge of the component

It is always possible to measure the thickness of the specimen at a free edge. The results may be determined with Vernier callipers or a micrometer. For laminates $\frac{1}{10}$ mm precision is acceptable. The readings may have to be compared with a standard or with the maximum and minimum thicknesses allowed in the specification.

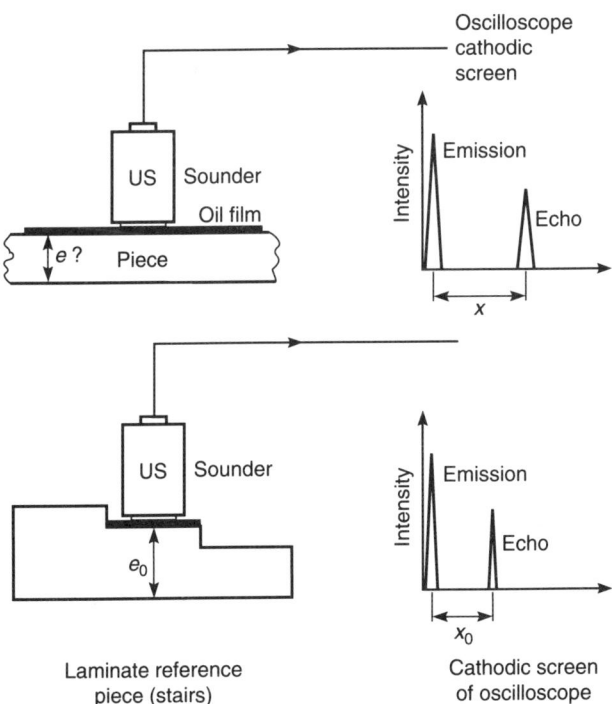

Figure 5.76 US thickness test.

(b) Test carried out on the centre of the component

Access to the centre can usually only be from one side. Hence the thickness is measured with ultrasonics, as shown in Figure 5.76. The reflection method is used to measure the thickness of a laminate (a reference unit or tiered or step shaped should be used to calibrate the instruments). The reference unit must be made of the same material and in the same way as the component being tested.

$$e = \frac{x}{x_0} e_0,$$

where x is proportional to e for the component and x_0 is proportional to e_0 for the reference unit. The step whose height e_0 is closest to the thickness of the component should be chosen for calibration.

This method can be used to measure the laminated skin in both monolithic laminates and sandwich structures.

5.3.4 Checking the shape of the component

(a) Principal defects relative to shape

These are defined by standard NFE 04-552. The principal defects referred to affect the shape of objects, which may be plane, straight, circular or cylindrical.

If the piece is plane, the symbol on the blueprints will be as shown in Figure 5.77, where X mm is the maximum distance between any point on the surface and the highest point.

If the piece is straight, the symbol on the blueprints will be as in Figure 5.78, where X mm is the largest admissible diameter of the cylinder determined by the length to be checked.

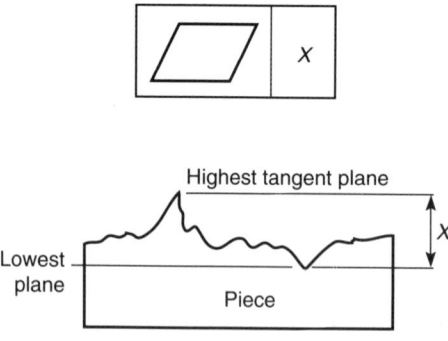

Figure 5.77 Plane component.

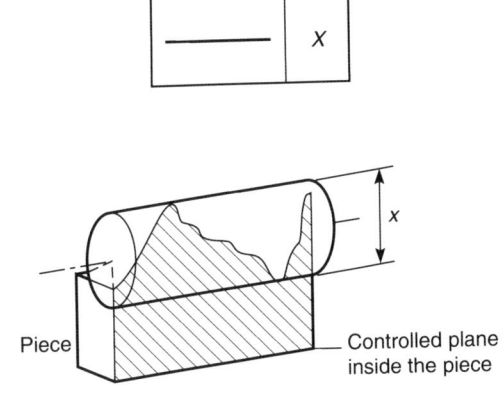

Figure 5.78 Straight component.

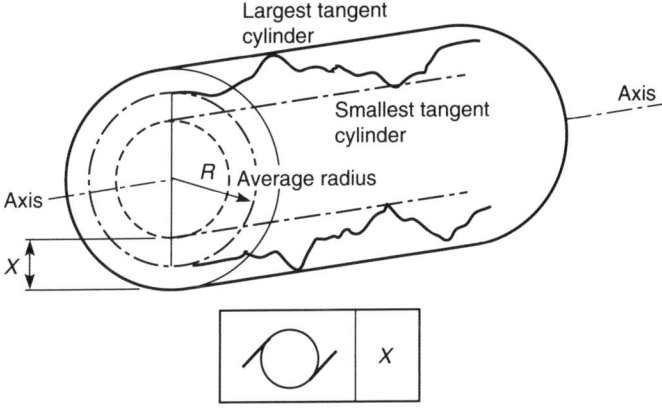

Figure 5.79 Cylindrical component.

If the piece is cylindrical, the symbol on the plan will be as in Figure 5.79, where X mm is the largest admissible radial distance between the lateral surface of the cylindrical piece undergoing testing and the largest diameter.

If the piece is circular, the symbol on the blueprints will be as in Figure 5.80.

(b) Principal checks

They can be inferred from the definitions given above. What must be checked is whether:

- a rod is straight;

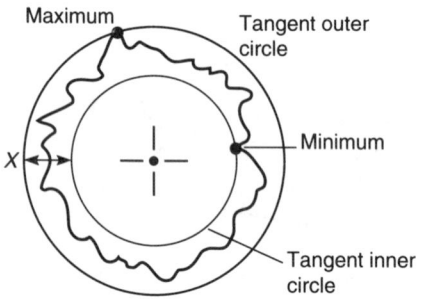

Figure 5.80 Circular component.

- a plate is flat;
- a cylinder is cylindrical.

Such tests are very important, and must be performed on demoulded, free, components.

(c) Causes of deformation

Components are mainly deformed because lamination has not been carried out symmetrically relative to the centre plane of the component. In other words the golden rule of fabrication: *lamination must be symmetrical about the central plane,* has not been observed.

When lamination is carried out at room temperature, the resin impregnated sheets are pliable enough to fit into the mould without being forced.

In the case of hot curing, the thermal balance between the mould and the piece is easily obtained. Dilatation takes place without any deformation or warping taking place.

When a unidirectional laminate cools down to room temperature, the individual lamina contract by different amounts in lengthwise and crosswise directions because of the orientation of the fibres. The laminate being briddled in the mould, contraction can cause deformations to take place. Consequently thermal stresses can occur within the components.

If the laminate is not a unidirectional one the layers may still contract anistropically according to the way they have been reinforced (type of weave), arranged and laid up.

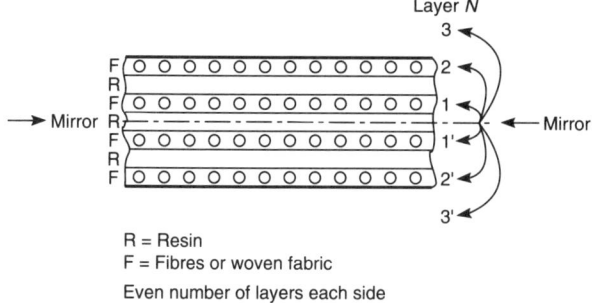

R = Resin
F = Fibres or woven fabric
Even number of layers each side

Figure 5.81 Mirror symmetry.

Once the piece has been demoulded, built-in stresses cause deformation and its shape is no longer similar to that of the mould. If the woven fabric or layers of material are laid in a symmetric manner on either side of the central plane, thermal stresses still occur. However, because of the symmetry of the construction, warping and distortion should not occur, though the stresses are still present (see Figure 5.81).

(d) Equipment used in testing

The equipment used in shape control tests is the same as that used in metrology:

- reference test bench;
- comparators;
- levels;
- optical rulers;
- laser beam;
- gauges and control assemblies;
- multi-axis measuring equipment in an air-conditioned room.

5.3.5 Checking surface positions and orientations

(a) Method

This consists of checking the position of surface S_1 in relation to surface S_0; S_0 and S_1 being the two sides of the same component, as shown in Figure 5.82. The surface S_0 is carefully defined and located in relation to the reference plane (bench). S_0 will preferably be one of the simple areas of the artefact, for instance a plane.

If the defects of surface area S_1 cannot be ignored, every single point, P_1, of S_1 must be checked. This can only be done rapidly if an automatic device linked to a computer is used.

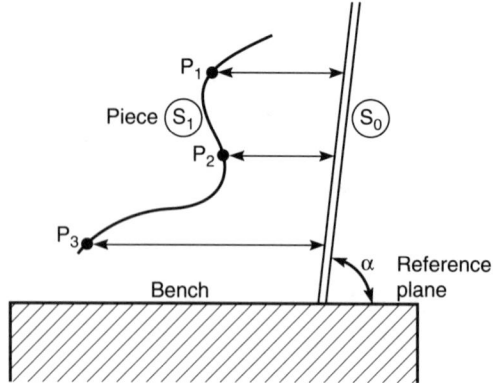

Figure 5.82

Figure 5.83

Position S_0/bench is known. Position of S_1/S_0 is measured. If the shape defects of surface area S_0 are negligible, it is sufficient to check the position of a few points, P_i, in order to determine the position of surface S_1 in relation to surface S_0.

(c) Simple tests

These include checking whether a plane is perpendicular to an axis and checking whether two cylinders are coaxial by rotating the piece around the measuring device or vice versa (see Figure 5.83).

5.3.6 Checking for warped surfaces (standard NFT 04.552)

Warping or flapping

Tolerance in the case of simple warping is the maximum variation X mm in the position of the surface S when it rotates around a reference axis (without

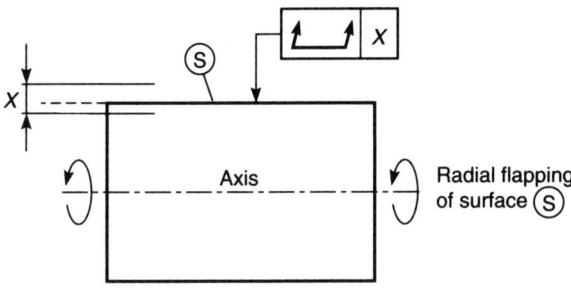

Figure 5.84

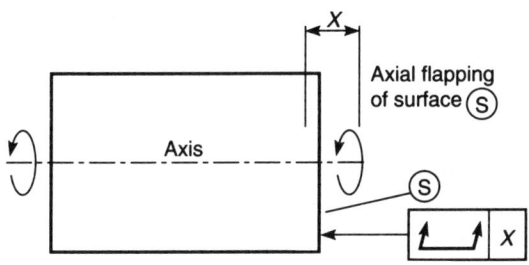

Figure 5.85

its being allowed to slide along this axis). Two kinds of flapping may be distinguished: radial and axial, as shown in Figures 5.84 and 5.85.

When radial warping occurs, it indicates that the component is not circular or cylindrical. The surface S should remain between two coaxial cylinders, X mm apart. X is the axial warping of surface S.

When axial warping occurs, it indicates that the piece is warped. The surface S must remain between two planes perpendicular to the axis and X mm apart.

5.3.7 Roughness of the surface

The surface roughness of composites may be due to the type of woven fabric used on the surface. This roughness is negligible, and there is no reason to measure it. Some components may be covered with a smooth resin gel coat for mechanical or cosmetic purposes. Any roughness of this will be caused by the mould itself. The real profile of any surface will probably show ridges and grooves as well as scratches, as shown in Figure 5.86, all of which should be measured.

Figure 5.86

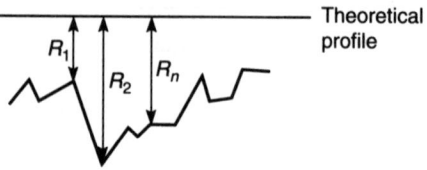

Figure 5.87

The quality of the surface can be characterized by R, which is the average depth of the defects, and R_a, which is the average difference relative to the mean surface elevation.

$$R_a = \frac{1}{l}\int_0^l |y|\,\mathrm{d}l \quad \text{and} \quad R = \frac{\Sigma R_i}{n},$$

where R_i is the distance between a groove and the adjacent ridge, as in Figure 5.87, and R_a is the arithmetical average of the absolute value of y selected with respect to the average surface elevation.

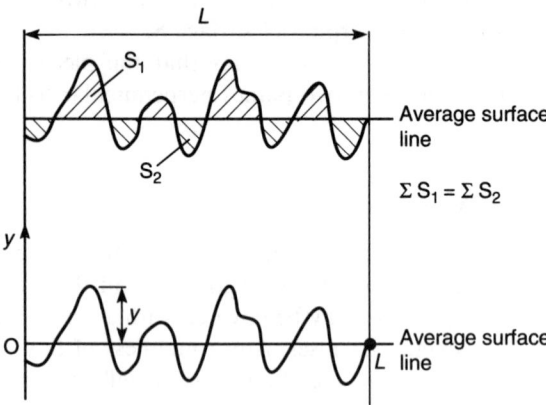

Figure 5.88

The values of R and R_a are expressed in microns and are measured with an electronic sensor. The sensor is run over the surface following the actual profile of the artefact, as shown in Figure 5.88. The vertical movements of the sensor are recorded. The profile of the test piece can be displayed on a printout and R and R_a computed.

5.3.8 Checking surface appearance

Composite parts for car bodies or consumer goods (electric appliances, etc.) must have:

- a glossy class A appearance;
- a uniform colour if they have been spun-dyed or coated with a gel coat.

The following controls are required:

- surface appearance and gloss control;
- dye control.

(a) General surface appearance control

This can be assessed:

- mechanically by using a profile sensor;
- optically either by the unaided eye or by measurements based on reflected or diffracted light.

The Daimler–Benz surface appearance analyser

A digital profile sensor, fitted with a needle, moves in a straight line along a guide rail in a step-by-step fashion. The values of the measured heights f_i are processed by a computer and can be read on a cutaway Oxy on a graph. These values are compared with the previously determined ideal curve. For each interval L, a standard deviation S and a wavelength λ of the profile can be inferred, as in Figure 5.89.

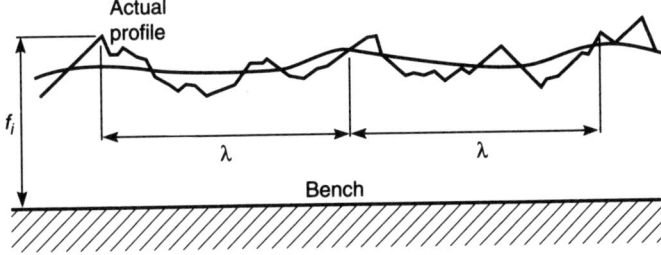

Figure 5.89

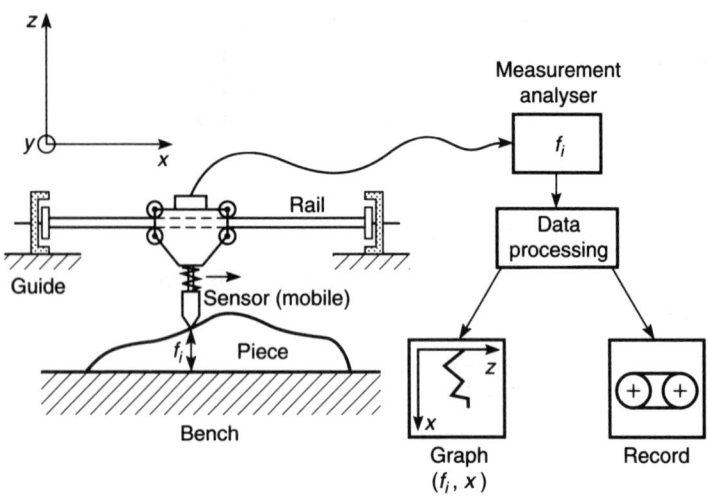

Figure 5.90 Surface appearance mechanical analyser (source: Daimler–Benz, Germany).

This task is rather time-consuming, taking about an hour to deal with a $1000\,\mathrm{cm}^2$ area. The device, which must be used in an air conditioned, vibration-free room, is really a piece of laboratory equipment and is illustrated in Figure 5.90. It is used by Daimler–Benz (Germany) and Union-Carbide (USA). The maximum size of the area under examination is:

$$L = 100\,\mathrm{cm} \qquad f_{\mathrm{max}} = 3\,\mathrm{cm}.$$

Optical control devices

Optical reflectometry. A fine grid is applied to the surface to be checked. The surface is illuminated and the incident light beam is reflected by the component on to a screen, as shown in Figure 5.91. The distortions that occur on the screen where the grid no longer appears regular because of the influence of defects (punctures, marks, undulations) are analysed. This method is effective, inexpensive and rapid; however, it does not provide quantitive output and needs experience to interpret.

Laser reflectometry. Loria®, the device developed by Ashland Chemicals, is based on the reflection of a laser beam from the central part of the piece to be checked, as shown in Figure 5.92. The composite which is to be examined must be conditioned (at 23 °C, 50% RH) beforehand. The device is installed in an air conditioned room. The laser beam scans the surface of the component along parallel lines, about 1 cm apart.

This method is accurate and fast (it only takes a few minutes to carry out the test) and allows surface quality to be quantified through the 'Loria index'. The lower the Loria index, the better the surface appearance of the material.

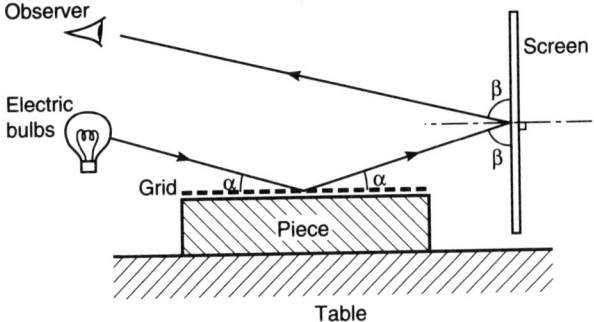

Figure 5.91 Optical reflectometry.

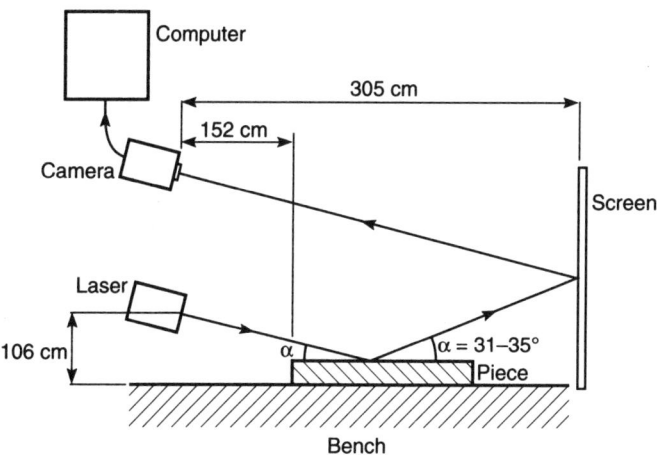

Figure 5.92 Laser reflectometry (source: Owens–Corning Fiberglass, USA).

Prior calibration is necessary. Plates are used for this purpose, each representative of a typical surface appearance.

The Loria index of SMC composite used to make a car body unit is ~70. The Loria apparatus is useful both for laboratory research and statistical quality control of surface appearance in the workshop. It is used by Owens–Corning Fiberglass (USA).

Optical diffractometry. The D-Sight® apparatus developed by Diffracto Ltd is based on the differentiation of light from the component and its reflection on a cylindrical screen and is illustrated in Figure 5.93.

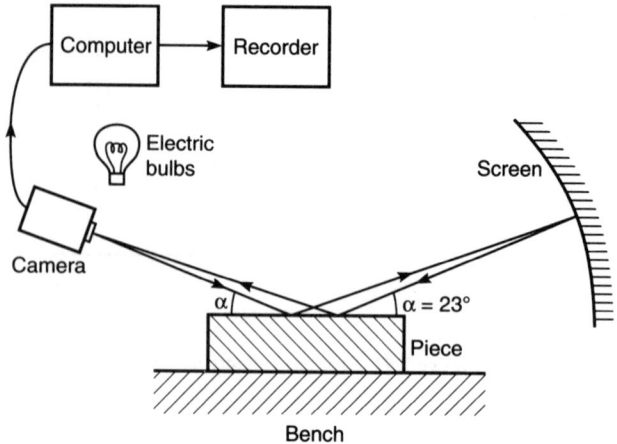

Figure 5.93 Optical diffractometry 'D-sight.'

The surface to be tested is lit twice, first by electric bulbs and second by light reflected from the screen. The angle of reflection of the light beams changes when there are defects on the surface. The picture recorded by the camera shows light and dark areas and it is possible to discriminate between defective areas (blisters, holes, cavities, undulations) and areas which have a perfect finish.

The events are recorded, stored, improved and processed by a computer. The final results are:

- the position (X, Y) of the defect on the surface;
- the size of the defect;
- the wavelength λ of undulations;
- the standard variation of the profile with reference to the theoretical profile.

The 'D-sight' apparatus works in real time, requiring 5–10 minutes for a measurement. It is well adapted to the statistical quality control of surface appearance and is employed by Fiat in Italy.

(b) Colour control

Purpose
Control by colorimetry enables gloss and colour to be measured by using the reflection of light. This technique is commonly used by paint manufacturers and can be applied to the control of gel coats and the surfaces of spun-dyed composite materials. Test pieces must be cut off the component and placed in the apparatus.

Different types of colorimeters
The following are available on the market.

- Manual optical colorimeters which can be used easily. The test piece is viewed through an eyepiece and compared with a standard reference specimen made up of superimposed yellow and blue filters. An example is the Lovibond® tinctometer.
- Automatic colorimeters fitted with microprocessors. Example: Sodexim (France).
- Spectrophotometers fitted with an integration sphere. Examples: Hunterlab® D25-P9, Philips SP8-100.

Spectrophotometer used with a microprocessor: principle
A halogen light source calibrated to give a suitable colour temperature is used to illuminate the surface. This beam of light passes through an IR filter to avoid heating the coloured surface. It then passes through a series of condensing lenses in order to produce a beam of coherent light which enters a 200 mm diameter sphere called an integration sphere. The beam of coherent light goes through the centre of the sphere and converges towards an outlet exactly opposite the inlet. The coloured test piece is placed at this outlet, as shown in Figure 5.94.

The incident beam is reflected by the coloured test piece. Diffracted light passes into the sphere which is coated with white barium sulphate to give a stable surface and provide a high spectral reflection curve. Only the diffracted light reaches the filters which are tangent to the upper part of the sphere. A photosensor is placed behind each filter. The sphere is fitted with four tangent filters and four silicon cells.

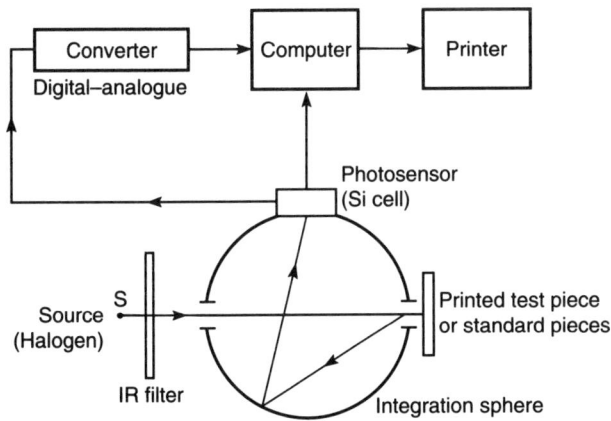

Figure 5.94 Spectrophotometer (colorimeter).

Thus it is possible to measure the light that is diffracted. The part of the beam which is reflected exactly opposite to the inlet by the test piece passes out again through the inlet and is therefore not taken into account. By slightly pivoting the test piece, it is possible to integrate the specular component into the measurement. Light diffraction is measured both on the coloured test piece and on standard black and white reference pieces. Comparison is made by computer.

The colorimetric values are measured relative to the absolute values given by the white standard reference piece (standard ASTM E306). The whole system has been designed to simulate the response of human colour vision. The light that is reflected accumulates in the sphere and is then diffracted by the white coat of barium sulphate. The photosensors produce electric signals which are proportional to the quantity of light received.

Control of finished product composite sandwich structures

6.1 PROCESSES USED TO MANUFACTURE THE PRODUCT

There are two ways of manufacturing sandwich structures: either in a single operation or in a two-stage process.

6.1.1 Two-stage process: lamination of skins plus adhesion of skins to the core

In this case the skins are monolithic laminates whose production must be controlled separately according to methods described in Chapter 5. Once the skins have been bonded on to the core, the complete assembly will have to be checked. Prior to this, the adhesive must be chosen and its quality ascertained before use. Surface treatment of the core and skins may be necessary before bonding. The curing cycle for the assembly process must also be checked.

The last two points are dealt with in Sections 4.2 and 2.6.

6.1.2 One-stage process: the skins are cured and bonded to the core in one operation

In this case the finished product control relates to both skin fabrication and skin–core bonding.

This method also requires that the adhesive be checked beforehand. The adhesive is either in the form of a film or in the form of a liquid or a paste.

The same adhesive may be used to fabricate the skins and to form the skin–core bond. The cure cycle will be as specified by the supplier.

6.2 LIST OF CONTROLS TO BE CARRIED OUT ON BONDED ASSEMBLIES

The quality of the bonding of a skin on to a sandwich core must be controlled while manufacturing the product. The following are required.

Checking the adhesive before use to verify that it is fresh. The product will have been checked upon delivery and the two sets of results should be compared.

Controlling the surfaces to be bonded. This means:

- checking the surface treatment bath (anodization) in which the aluminium honeycomb is treated;
- checking the way the surface treatment has been performed.

Checking the handling of components and adhesive: gloves should be worn for personal protection and to avoid contaminating surfaces with grease, etc.

Checking the spreading of the adhesive on to the sandwich core.

- If the adhesive is a paste, a notched spatula should be used.
- If the adhesive is a liquid or a resin, the surface to be bonded should be given a good coat of adhesive, and care taken to minimize bubbles (possibly by rolling).
- If the adhesive is a film, the protective surface films should be removed. They are coloured to make detection easier.

Controlling the curing cycle of the assembly.

- Special cycle for an adhesive paste or film.
- Normal curing cycle of the resin if it is used to bond the skin on to the core as well as form the skins (this is co-curing).

6.3. DESTRUCTIVE TESTS ON SANDWICH STRUCTURES

It is not usually feasible to remove a sample of the final product for testing because of the expensive nature of these composites. However when a long series of comparatively cheap sandwich pieces is produced, it may be possible to set one aside for testing and evaluation. Alternatively, when larger pieces are produced, an extra length can be made for test purposes.

6.3.1 Tests on the skins

If the skin can be removed by peeling off, or the core removed by machining, without damaging the skin, it is possible to carry out the set of destructive tests described in Section 5.1, that is:

- fibre and resin ratio;

- density;
- pore content;
- state of cure;
- mechanical properties.

These tests should be carried for each batch of skins especially if the skins are fabricated separately from the complete structure.

6.3.2 Tests on the core

These are carried out on the as-received material (see Section 2.5).

6.3.3 Testing skin–core bonding

Inadequate skin–core bonding may come from lack of or too much adhesive, too many bubbles or voids in the bonding resin or incomplete cure of the adhesive.

Bond quality can be assessed by a tensile test perpendicular to the adhesive layer or a peel test.

(a) Resistance to transverse tension
(Standard NF L17-452, no ISO standard)

This standard was developed to assess the adherence of the laminated skin to the honeycomb core, but it can be used with any other kind of core (rigid foam, balsa wood). A square ($S = 50 \times 50\,\text{mm}^2$) is cut off from the sandwich structure. If the core is a honeycomb as in Figure 6.1, two of the sides must be parallel to the length (L) of the cells. Two ASTM 2017 duralumin blocks are glued on to the skins of the test piece with a good adhesive.

Selecting the latter may present difficulty. To avoid damaging or changing the specimen, it would be advisable to use a strong cold epoxy cure adhesive.

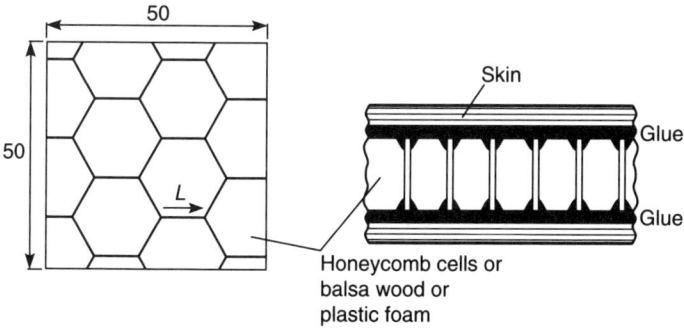

Figure 6.1 Sandwich test piece.

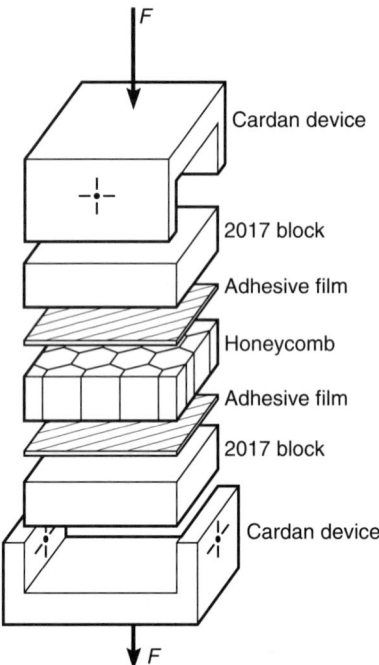

Figure 6.2 Honeycomb duralumin traction test.

Unfortunately, cold cure resins do not possess as good a mechanical performance as those cured at temperature. Hence, the standard recommends that a medium-temperature cure adhesive should be used provided its curing temperature is compatible with the thermal resistance of the laminated skin. In practice a system with a curing temperature less than that of the adhesive used in the skin–core bond should be employed. See Figures 6.2 and 6.3. Just before bonding, the blocks and the sandwich specimen should be abraded with fine grit paper and cleaned with a degreasing agent (ketone).

The test piece is mounted in a jig which automatically ensures that the load is perpendicular to the block. It is then strained slowly (speed: 1 mm/min) until it breaks under a load, Fr. The adhesive stress between the skin and the core is:

$$\sigma = \frac{\text{Fr}}{S}$$

where S is the surface area.
Test pieces which break in any of the following modes are rejected:

- in the core of the sandwich structure;
- because of adhesive failure between the skin and block;
- in the thickness of the laminated skin.

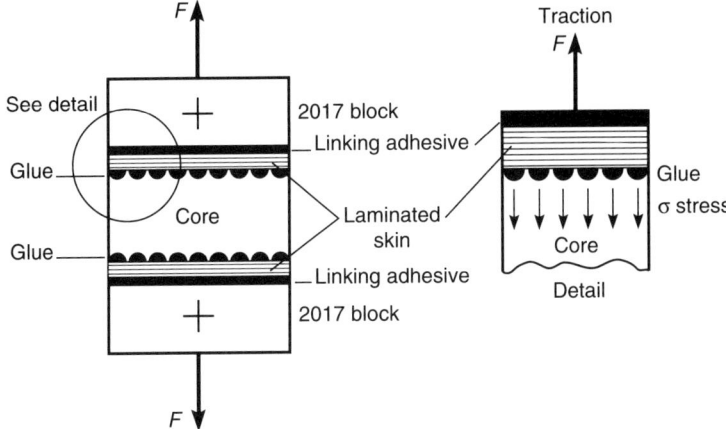

Figure 6.3 Flatwise tensile test.

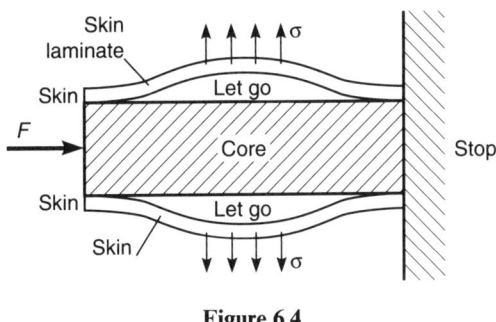

Figure 6.4

The test is repeated on several samples. This test gives useful information on the capacity of the adhesive to withstand local buckling. When a sandwich structure is subjected to pressure on one of its ends (see Figure 6.4), the skins are apt to buckle, i.e. to debond from the core.

(b) Peeling test
(NF L17-455) (no ISO standard)

Peeling occurs when the resistance of adhesives is inadequate; the skin literally peels off. The principle is shown in Figure 6.5. The forces acting are complex and the results cannot be interpreted simply.

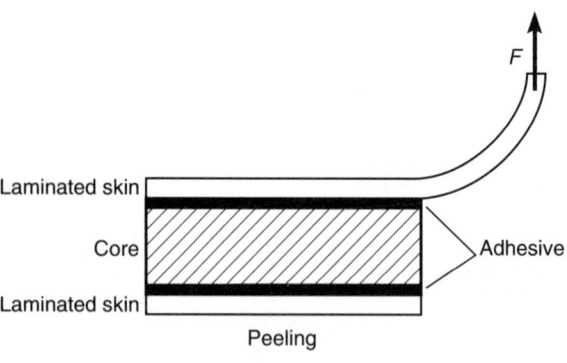

Figure 6.5

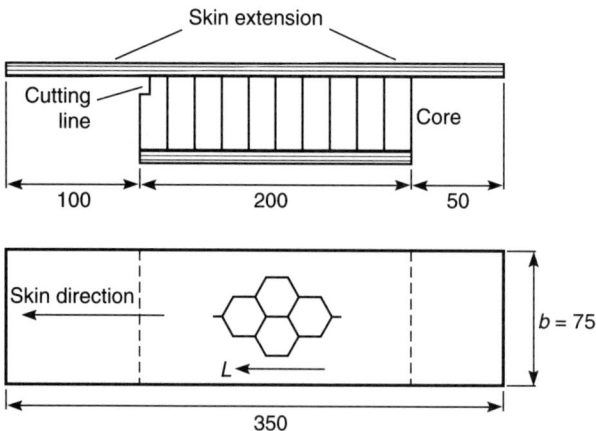

Figure 6.6 Peeling test piece.

Though the standard was developed to assess the adherence of the skin to a honeycomb core, it can be used with any other type of core (foam, balsa wood, etc.). The test piece is a rectangle cut off from the sandwich structure as shown in Figure 6.6. Part of one of the skins extends beyond the core to allow the stressing force to be applied.

If the core is made of honeycomb the cells should be parallel to the length L of the test piece, i.e. in the direction in which peeling will occur. A saw cut in the core of the sandwich structure will initiate the peeling process, which is shown in Figure 6.7.

The torque (C_p) needed to detach the skin and wind it around the drum is measured during the test. The peeling force (F_p) is expressed in newtons

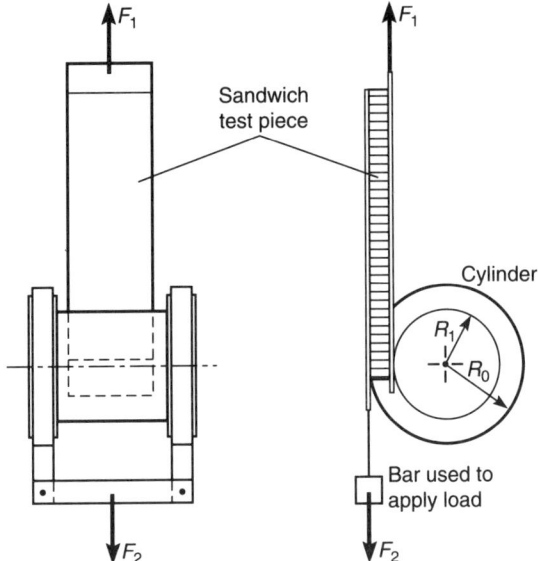

Figure 6.7 Peeling test.

and the resistance of the skin to peeling is the torque (C_p) per unit of width of the test piece. The test is carried out on a minimum of five specimens. The drum diameter $D_1 = 100$ mm and the skin is removed at a constant speed of 25 mm/min. The variation of force F during the course of winding is recorded as in Figure 6.8. It is irregular because of the stick–slip nature of the failure process. Fr is the average force applied; F_0 is the force required to move the drum; $F_p = \mathrm{Fr} - F_0$ is the peeling force and $C_p = F_p \times (R_0 - R_1)/b$ is the peeling torque per unit width. Any tests in which the skin breaks or fails by delamination are discounted.

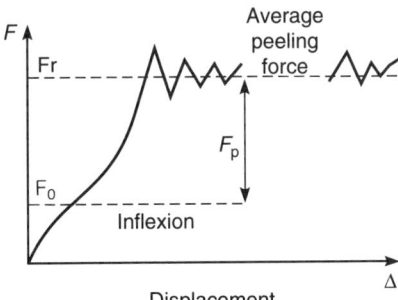

Figure 6.8

6.3.4 Flexural and compression tests on sandwich structures

Sandwich structures are mainly used for their stiffness, i.e. as beams or plates. The following quality control tests are often employed:

- three-point flexural test;
- compression/buckling test.

No standards have been developed so far to describe the test procedures.

(a) Flexural test

This test has already been described in Chapter 5 for a plane laminate. The principle, which is the same for a sandwich structure, is illustrated in Figure 6.9.

The approximate tensile strength of the lower skin is:

$$R = \frac{3F_m \cdot D}{2be^2},$$

where D is the distance between support points, b the breadth and e the thickness of the specimen. The overall flexural strength of the structure is assumed to be the same as the above.

(b) Compression/buckling test

A sample of width b and thickness e is cut from the structure and mounted with an adhesive between two blocks of 2017 or 2024 (ASTM) such that the skins are vertical. It is then slowly loaded in compression, as shown in Figure 6.10.

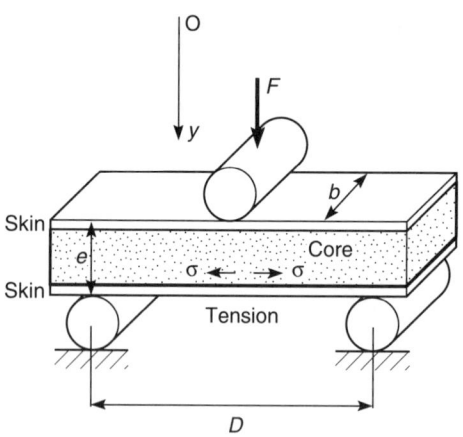

Figure 6.9 Flexural test.

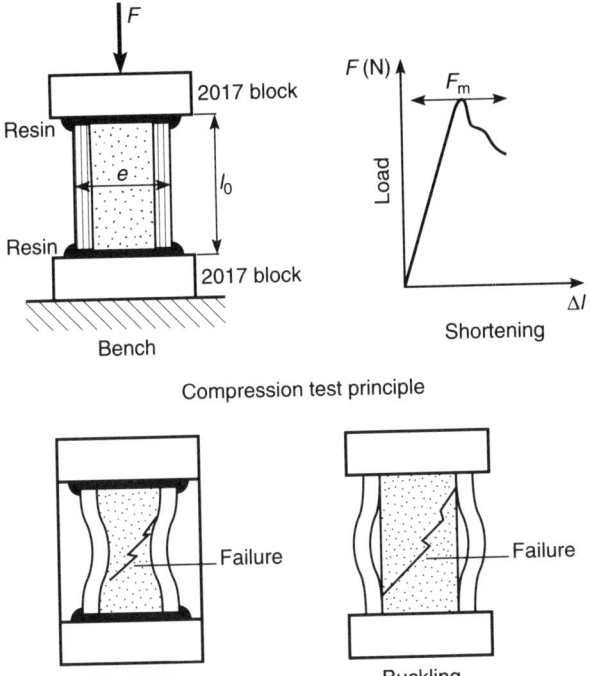

Figure 6.10 Results of compression test.

Failure may be by buckling or gross structural fracture. The resistance of the sandwich structure to compression/buckling is:

$$R = \frac{F_m}{b \times e}.$$

It should be noted that this is an overall value. Stress gauges on the skins indicate very high stress values in the vicinity of the buckling or failure point.

According to Gay (1987), the critical local buckling stress in the skin sandwich structure is:

$$\sigma_{crit} = 3\{12(3 - v_c)^2 \times (1 + v_c)^2\}^{1/3} \times \{E_p \times E_c^2\}^{1/3},$$

where E_p is the stiffness modulus of the skin, E_c is the stiffness modulus of the core and v_c is Poisson's ratio of the core.

6.4 NON-DESTRUCTIVE TESTS ON SANDWICH STRUCTURES

The test is carried out on the component itself. It does not provide any information on the quality of the bond, but it indicates whether bonding has occurred or not. It also indicates whether delamination has taken place in the skins. The following tests may be used.

6.4.1 Sonic tapping tests

When the component is tapped lightly with a small hammer, the various sounds produced indicate the areas where there is no bonding as well as the places where delamination has occurred. A special hammer fitted with an accelerometer may be used for this purpose.

6.4.2 US test

An ultrasonic emitter is moved across the surface of the sandwich structure and the transmitted signal detected. A coupling liquid (oil, water, or a simple adhesive paste) is used to ensure good contact between the transducer and the sandwich structure. Two types of defect may be noted:

- an adhesive bond failure at the skin–core interface of the structure;
- delamination within the skin.

Defect detection is facilitated by calibrating the apparatus on a sandwich structure known to be defective. See Figure 6.11.

6.4.3 Resonance test

For example with the Fokker bond tester, there is a correlation between the resonance frequency of a bond and its degree of adhesion, hence its resistance to failure. This resonance frequency is compared with that of a defect-free reference sample.

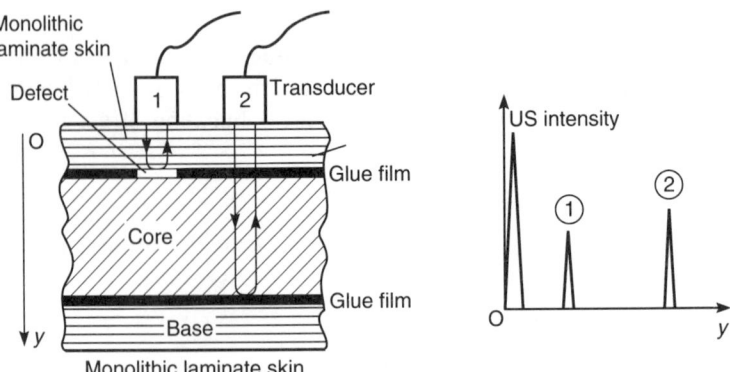

Figure 6.11 US test ① is the echo from the defect; ② is the echo from the base or skin on the other side.

6.4.4 X-ray test

This test can detect the adhesive film used at the interface if the absorption coefficient of the rest of the specimen is smaller than or at most equal to that of the adhesive.

6.4.5 X-ray scanning tomography see Section 5.2.11

6.4.6 IR thermography test

The component is heated evenly. Any defects present in the bond will disrupt the flow of heat through the specimen. This is a method used to study the whole unit. The defective areas may be easily found, as in Figure 6.12.

6.4.7 Laser holograph interferometry

The component is stressed and hence slightly deformed, and a laser beam focused on to it. The areas which contain defects (the absence of a bond or a defective bond) differ in appearance from the defect-free material because of non-uniform deformation.The test is costly to perform.

6.4.8 Neutron radiography testing

This approach is used to complement X-ray testing for materials which have very similar constituents, with close atomic numbers and similar densities. This is the case for composite reinforcements and resins used for lamination or bonding.

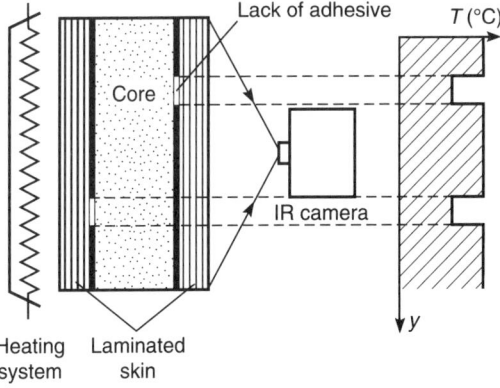

Figure 6.12 IR thermography.

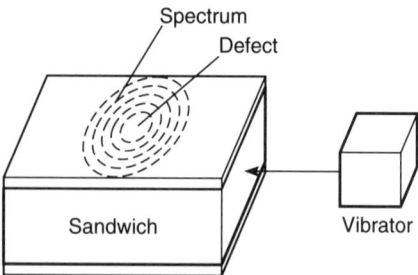

Figure 6.13 Vibration spectrum.

It is thus possible to pick out the lack of homogeneity in the bond, the differences in the thickness of the glue line and the lack of adhesive.

Neutron radiography requires the availability of a neutron source which has to be supplied from a specially adapted nuclear reactor. Consequently the technique tends to be costly and is rarely available close to the fabrication facility.

6.4.9 Other methods

Bonds can be checked, depending on their nature and shape, by their vibration spectrum. This method is somewhat insensitive. It can, however, be used to test plane plates.

Better alternatives are available, such as testing for specific resonance modes or holographic interferometry. The vibration spectrum method consists of spreading very fine coloured sand on the plane plate and then exciting vibrations in the structure. Defects alter the local vibration behaviour, as shown in Figure 6.13.

Another method is checking the vacuum tightness of the bond between the skin and core. The test used depends on the size of the leak that is suspected. Methods employed are soap bubbles, helium or ammonia (NH_3).

Conclusions | 7

The composite fabrication industry is gradually moving towards large-scale production. However, production is presently very labour intensive and the results depend on the quality and skill of the workforce to a greater degree than in more highly automated industries. Hence proper quality control is even more important for composite manufacture than for other industries.

Owens–Corning Fiberglass conducted a survey in 1987 which showed that 82% of the companies polled believed quality was the one factor that the composite fabrication industry must emphasize if composite artefacts were to remain superior to those made from competing materials. For many small companies which produce glass fibre resin composites by contact moulding or by using room temperature curing resins, any improvement in quality will be achieved by resorting to simple, inexpensive procedures such as:

- storing resins in a cool place at a constant temperature;
- verifying the curing characteristics of the resin;
- adequately removing air bubbles from the laminate;
- measuring Barcol or Shore hardness;
- checking the glass temperature T_g;
- checking surface appearance;
- checking dimensions after curing;
- ensuring that the workshop is adequately ventilated and heated to avoid excessive humidity.

Composites are complex materials which require numerous raw materials and manufacturing techniques to be used.

It is not surprising that many quality control measurements are necessary to obtain good-quality products. There is a danger that after reading this manual, manufacturers might feel there are too many quality control or assurance techniques and be put off using them because of the high cost of testing. In fact, the testing methods described in this book, and there are more of them, are not all required in given circumstances. A producer of composite products is free to choose the quality controls that s/he believes to be the cheapest and best adapted to his/her product. Quality control is expensive, but having products rejected because they are defective is even more costly! Quality control accounts for 20% of the total cost of a satellite structure.

It is likely that within the next few years, certain tests will come to the fore as the preferred and best value for money. This will occur naturally, as has been the case with tests for metal artefacts. Be that as it may, quality control is and will remain essential in order to reduce the number of rejected products (and thus bring down manufacturing costs) and to build up customer confidence in composite materials. That is the price that must be paid if the buyers of consumer goods are to select composite products in preference to conventional ones. *Quality is the condition required for the expansion of the composite market.*

References 8

Normes AFNOR Volumes 1, 2 and 8 (1986–1991), France.

Chrétien, G. (1986) *Materiaux Composites à Matrice Organique*, Lavoisier, France.

Gay, D. (1987) *Materiaux Composites*, Hermes, France.

Documentation Cetim, Nantes, France.

Geier, M. (1985) *Guide Pratique des Materiaux Composites*, Lavoisier, France.

Weiss-Bord (1983) *Les Materiaux Composites*, Edition L'Usine Nouvelle, France.

9 | Some useful addresses for equipment

Appliance and trademark	Name of supplier	Address
VISCOMETERS		
Gel Norm	O.S.I.	141 rue de Javel 75015 PARIS (France)
Trombomat	Prodemat	Centre des Samourais 2 allée Clouzot 69100 VILLEURBANNE (France)
CURING SIMULATORS		
Kinemat	Prodemat	(see above)
Plastoreactomat	Prodemat	(see above)
Vanhographe	Cotton frères	88 boulevard des Belges 69006 LYON (France)
STICKOMETERS		
Pegosimetre	Gira	Rue des Bruyères ZI Morlass-Berlanne 64160 MORLASS-CEDEX 21 (France)
SCANNERS		
Tomographe Industriel	Intercontrole	13 rue du Capricorne 94583 RUNGIS (France)
DIFFERENTIAL MECHANICAL ANALYSER	Delta T.A. Instruments	11 boulevard de Lyon 59100 ROUBAIX (France)
DMA	(Du Pont de Nemours)	
DIFFERENTIAL SCANNING ANALYSER DSC TA 3000	Mettler Toledo Instruments	18–20 avenue de la Pépinière ZAE BP14/78220 VIROFLAY (France)

Appliance and trademark	Name of supplier	Address
THERMOGRAPH IR Thermograph analyzer	Cedip	Cité Descartes 1 rue A Einstein 77436 MARNE LA VALLEE (France)
Camera AGA-VISION 870	Agema	18 rue Hoche BP81 92134 ISSY LES MOULINEAUX (France)
COLORIMETERS	Sodexim SA	ZI rue Nouvelle Jonchery sur Vesle 51140 MUIZON (France)
	Hunter Lab	114 avenue du Pt Wilson 93212 LA PLAINE ST DENIS (France)
CONTROL WITH **US WAVES** Epoch 2	Sofranel	59 rue Parmentier 78500 SARTROUVILLE (France)
ACOUSTIC TEST	Euro-Physical Acoustic	74 rue des Grands Champs 75020 PARIS (France)
TAP TEST Microphones	Metravib	565 rue des Sans-soucis 69760 LIMONET (France)
CONTROL WITH **EDDY CURRENT** DCF 500A	Ardrox SA	41 rue des Francs-Bourgeois 75004 PARIS (France)
DYE PENETRATION **TEST**	Ardrox SA	(see above)

Index

Page numbers appearing in **bold** refer to figures and page numbers appearing in *italic* refer to tables.